PRAISE FOR *A FEAST OF SCIENCE*

"Huzzah! Dr. Joe does it again! Another masterwork of demarcating non-science from science and more generally nonsense from sense. The world needs his discernment."

— Dr. Brian Alters, Professor, Chapman University

PRAISE FOR *THE FLY IN THE OINTMENT*

"Joe Schwarcz has done it again. In fact, he has outdone it. This book is every bit as entertaining, informative, and authoritative as his previous celebrated collections, but contains enriched social fiber and 10 percent more attitude per chapter. Whether he's assessing the legacy of Rachel Carson, coping with penile underachievement in alligators, or revealing the curdling secrets of cheese, Schwarcz never fails to fascinate."

— Curt Supplee, former science editor, *Washington Post*

PRAISE FOR *DR. JOE AND WHAT YOU DIDN'T KNOW*

"Any science writer can come up with the answers. But only Dr. Joe can turn the world's most fascinating questions into a compelling journey through the great scientific mysteries of everyday life. *Dr. Joe and What You Didn't Know* proves yet again that all great science springs from the curiosity of asking the simple question . . . and that Dr. Joe is one of the great science storytellers with both all the questions and answers."

— Paul Lewis, president and general manager, Discovery Channel

PRAISE FOR *THAT'S THE WAY THE COOKIE CRUMBLES*

"Schwarcz explains science in such a calm, compelling manner, you can't help but heed his words. How else to explain why I'm now stir-frying cabbage for dinner and seeing its cruciferous cousins — broccoli, cauliflower, and brussels sprouts — in a delicious new light?"

— Cynthia David, *Toronto Star*

PRAISE FOR *RADAR, HULA HOOPS, AND PLAYFUL PIGS*

"It is hard to believe that anyone could be drawn to such a dull and smelly subject as chemistry — until, that is, one picks up Joe Schwarcz's book and is reminded that with every breath and feeling one is experiencing chemistry. Falling in love, we all know, is a matter of the right chemistry. Schwarcz gets his chemistry right, and hooks his readers."

— John C. Polanyi, Nobel Laureate

ALSO BY DR. JOE SCHWARCZ

*A Feast of Science: Intriguing Morsels from the
Science of Everyday Life*

*Monkeys, Myths, and Molecules: Separating Fact from Fiction,
and the Science of Everyday Life*

*Is That a Fact?: Frauds, Quacks,
and the Real Science of Everyday Life*

*The Right Chemistry: 108 Enlightening, Nutritious,
Health-Conscious, and Occasionally Bizarre Inquiries
into the Science of Everyday Life*

*Dr. Joe's Health Lab: 164 Amazing Insights into
the Science of Medicine, Nutrition, and Well-Being*

*Dr. Joe's Brain Sparks: 179 Inspiring and Enlightening
Inquiries into the Science of Everyday Life*

*Dr. Joe's Science, Sense & Nonsense: 61 Nourishing, Healthy,
Bunk-Free Commentaries on the Chemistry That Affects Us All*

*Brain Fuel: 199 Mind-Expanding Inquiries
into the Science of Everyday Life*

*An Apple a Day: The Myths, Misconceptions,
and Truths About the Foods We Eat*

*Let Them Eat Flax: 70 All-New Commentaries
on the Science of Everyday Food & Life*

*The Fly in the Ointment: 70 Fascinating
Commentaries on the Science of Everyday Life*

*Dr. Joe and What You Didn't Know: 177 Fascinating Questions
and Answers About the Chemistry of Everyday Life*

*That's the Way the Cookie Crumbles: 62 All-New Commentaries
on the Fascinating Chemistry of Everyday Life*

*The Genie in the Bottle: 64 All-New Commentaries on the
Fascinating Chemistry of Everyday Life*

*Radar, Hula Hoops, and Playful Pigs: 67 Digestible Commentaries
on the Fascinating Chemistry of Everyday Life*

A GRAIN OF
SALT

*The Science and Pseudoscience
of What We Eat*

Dr. JOE SCHWARCZ

Published in Canada by ECW Press
665 Gerrard Street East
Toronto, Ontario, Canada M4M 1Y2
416-694-3348 / info@ecwpress.com

Cover design: David A. Gee

LIBRARY AND ARCHIVES CANADA
CATALOGUING IN PUBLICATION

Title: A grain of salt : the science and
pseudoscience of what we eat / Dr. Joe Schwarcz.

Names: Schwarcz, Joe, author.

Description: Includes index.

Identifiers: Canadiana (print) 2019010922X
Canadiana (ebook) 20190109238

ISBN 9781770414754 (softcover)
ISBN 9781773053868 (PDF)
ISBN 9781773053851 (ePUB)

Subjects: LCSH: Food—Miscellanea. | LCSH:
Diet—Miscellanea. | LCSH: Food habits—
Miscellanea. | LCSH: Food—Popular works. |
LCSH: Diet—Popular works. | LCSH: Food
habits—Popular works.

Classification: LCC TX355 .S39 2019
DDC 641.3002—dc23

The publication of *A Grain of Salt* has been funded in part by the Government of Canada. *Ce livre
est financé en partie par le gouvernement du Canada.* We also acknowledge the contribution of the
Government of Ontario through the Ontario Book Publishing Tax Credit, and through Ontario Creates
for the marketing of this book.

PRINTED AND BOUND IN CANADA

PRINTING: MARQUIS 5 4 3 2 1

INTRODUCTION

Whether it be after a public lecture or on my radio show or via email, questions I get often begin with "Is it true that . . . ?" Usually these questions are triggered by a blog post, something heard on the radio or seen on the *Dr. Oz Show*. Invariably my answer is, "Well, you'd better take that with a grain of salt." That's a pretty common expression encouraging a generous dose of skepticism.

This classic phrase supposedly originated with the Roman historian Pliny the Elder, who recounted how Pompey the Great found a note written by Mithridates, king of Pontus, after Pompey defeated him in battle. Mithridates was terrified of being poisoned and devised various supposed antidotes to use in case someone tried to do him in with some sort of toxic substance. The note that Pompey found described a concoction of two dried walnuts, two figs, and twenty leaves of rue. A grain of salt was to be added for palatability!

Taking this mixture every day was alleged to offer protection against all poisons. The unlikelihood of this being effective gave rise to the notion that whatever was being taken with a grain of salt warranted scrutiny. According to the popular legend, when Mithridates's armies were defeated by the Romans, he tried to commit suicide by taking poison but was unsuccessful because his "mithidratum" had made him immune! That story, of course, should also be taken with a grain of salt. As should much of the nutritional information the media floods us with.

Let me illustrate: recently, I was asked about a headline featured on the Facebook page of an organization that goes by the name The Hearty Soul. That page has over a million followers, so it can have considerable impact. The seductive headline was "Study: Eating Chocolate and Drinking Red Wine Could Help Prevent Aging." Visitors who didn't get beyond the headline, and I suspect most didn't, were likely to leave with the notion that researchers had confirmed that chocolate and red wine prevent aging. They may even have applied the findings to their own life. However, the study referenced in the article in no way showed that eating chocolate or drinking red wine prevents aging.

What the researchers did demonstrate was that when cells grown in tissue culture age, some of their youthful properties are restored upon exposure to chemicals called resveralogues. But rejuvenation of cells in a culture dish is a long way from preventing aging in humans. And what is the chocolate and wine connection? Both contain resveratrol, a compound that has been extensively studied for health benefits based on some intriguing preliminary findings in rodents. However, any benefits of resveratrol in mice only occur with doses that are far greater than those achievable by food or beverage intake. That's why researchers decided to play molecular roulette and synthesize compounds similar to resveratrol, the resveralogues, with hopes of seeing greater potency at lower doses.

The activity of these resveralogues was reported in a paper published in *BMC Cell Biology*, a very reputable medical journal. The findings are certainly of academic interest, but they have nothing to do with chocolate or red wine, neither of which contain the compounds tested. Nothing about this study demonstrates that eating chocolate or drinking wine counters aging, and the authors do not make that claim at all. But the sensationalized headlines do.

This story is a prime example of the need to interpret scientific studies for the public in a proper fashion and thereby hopefully prevent misinformation from taking flight. This is particularly challenging when it comes to food and the question of what we should or should not eat. Confusion abounds — not due to a lack of information. Quite the opposite. Virtually every day we are alerted to some new study that claims to add to our knowledge about pesticides, genetic modification, food additives, organic produce, probiotics, weight loss regimens, and an array of "toxins" that may be contaminating our food supply. Some raise fears, others allay them. The media often sensationalizes the findings, and bloggers with various levels of expertise weigh in, often twisting the data to suit a particular agenda. The public is left dismayed and bewildered about what and whom to believe.

What regimen should be followed? Gluten-free? GMO-free? Dairy-free? Organic? Vegan? Keto? Paleo? Mediterranean? Should lectins be avoided? Breakfast skipped? Plastic packaging shunned? Are probiotics worthwhile? How about antioxidants? Vitamin supplements? Omega-3 fats? Meat? Farmed fish? Eggs? Soy? Is alcohol good or bad? The questions seem endless. As do the answers.

Why, people wonder, with all the effort and money that has been spent on nutritional research over the years, is there still so much puzzlement about what we should eat? The answer to that question is actually simple. The human body is a staggeringly complex machine, and our food supply is a staggeringly complex mixture of chemicals, over 99.9 percent of which are naturally occurring. Predicting the outcome of blending these complexities is extremely challenging. Factor in the influence of genetics, food intolerances, allergies, biochemical individuality, and age, and you have a practically unsolvable problem when it comes to deciding exactly what we should eat.

Add to this the fact that most studies of human nutrition involve notoriously unreliable food frequency questionnaires. I suspect you would have a hard time answering questions about how many apples, how many servings of broccoli, or how many eggs you consumed over the last six months. It is hard to remember exactly what we ate yesterday, never mind over an extended period. Yet almost all nutritional studies are based on food questionnaires that rely on such recall. Furthermore, people are likely to overestimate foods that are thought to be healthy and underestimate supposedly unhealthy foods. They are unlikely to advertise their penchant for fried foods and seem to be so in love with broccoli that they apparently eat more than is actually produced.

Certainly, we can make some educated guesses about the risks and benefits of various components of our diets, and we can come up with general guidelines for healthy eating, but unfortunately science cannot always provide definitive answers. Pseudoscience, however, knows no such boundaries.

What are those basic guidelines? Eat lots of fruits, vegetables, and whole grains, emphasize mono and polyunsaturated fats while minimizing salt, sugar, refined grains, red meat, processed meats, soft drinks, and saturated and trans fats. This is backed up by a landmark twenty-seven-year global diet analysis published in the highly respected journal *The Lancet* in April of 2019. Thousands of researchers tracked premature deaths and disabilities associated with 350 diseases in 195 countries over a twenty-seven-year period in what may be the most extensive study ever on the relationship between diet and disease. The conclusion was that roughly one in five premature deaths can be attributed to diet, with a low intake of healthy foods being a greater contributor than a high intake of unhealthy foods. High sodium intake and a lack of whole grains, fruit, nuts, and seeds were the most important factors linked with disease, more than consumption of red and processed meats, trans fats, or sugary beverages, although these were still risk factors.

In the following pages, I do not propose to provide a comprehensive treatise on food and nutrition, nor do I claim to be able to demystify all the complexities of our diet and food supply. Rather, I have chosen some topics based on questions I have been asked, and others just because I find them interesting. Each entry is meant to stand on its own. So, do a little tasting, bite into one that strikes your fancy, and feed your appetite for knowledge. But let me remind you that in the area of food and nutrition, just about all information has to be taken with a grain of salt.

FRUITS OF THE INTERNET

When I graduated university, my parents gave me a very special gift: a set of the renowned *Encyclopedia Britannica*! It had the answer to virtually every question I came across and was fun to just browse through. The last time I picked up one of those heavy volumes was about twenty years ago. By then, the Internet and the tsunami of information it brought to our fingertips had appeared on the scene.

Every day, I witness both the positive and negative power the web can unleash. I'm continuously flooded with questions that can be traced back to some item seen on the web. You don't even have to watch television anymore because significant clips, or entire programs, are just a few keystrokes away. A glance at just a few of the hundred or so questions that come my way during a single week affords insight into what is on the public mind. So let's have a go.

I always know what Dr. Mehmet Oz has been up to because my email inbox boils over with questions about his latest antics. Judging by the number of questions I got about monk fruit, it was clear that Oz had been trying to sweeten people's lives with this alternative to artificial sweeteners. We love sweets, but we worry, justifiably, about consuming too much sugar, and less justifiably, about its artificial replacements. The market is ripe for products that can be promoted as "natural no calorie sweeteners." Monk fruit extracts happen to fit the bill.

Legend has it that the fruit, commonly known by its Chinese name, *luo han guo*, was first cultivated by Buddhist monks back in the thirteenth century for its supposed fever-reducing and cough-relieving properties. Folklore also has monk fruit extending life, with claims that the counties in China where the fruit is grown for commercial purposes have an unusual number of centenarians. This has never been confirmed, and neither do we know whether the fruit is routinely consumed by the population there.

Various preparations of the fruit are sold in China with claims of moistening the lungs, eliminating phlegm, stopping cough, relieving sunstroke, and promoting bowel movements. While the efficacy of monk fruit as a medicine is questionable, the sweetness of its juice is not. This, however, did not arouse scientists' curiosity until the 1970s, when expanding waistlines led to expanding markets for noncaloric sweeteners. Analysis revealed that monk fruit contained five closely related compounds called mogrosides that are some 250 to 400 times sweeter than sugar. These can be extracted from the juice and processed into a powder to be used as a sweetening agent. Because of the high degree of sweetness, very little of the monk fruit extract is required, so it can be labeled as noncaloric. Although there have been no extensive safety studies, the U.S. Food and Drug Administration has classified monk fruit preparations as generally recognized as safe, or GRAS.

McNeil Nutritionals has now introduced monk fruit extract as a "no calorie sweetener" under the name Nectresse. In order to provide volume and appropriate texture, the extract is blended with a small amount of erythritol (a sugar alcohol), sugar, and molasses. These contribute fewer than five calories per serving, which is the limit for a product to be labeled as "containing no calories." I'd have no problem with trying this sweetener, and I'm sure many people who worry about artificial sweeteners will pounce on it. And people are worried about this one, sometimes for unusual reasons.

I was asked whether it is true that aspartame is made from the waste of certain bacteria. Actually, it sort of is, but that is irrelevant. Enzymes churned out by bacteria are commonly used to produce chemicals. Just think of adding *Lactobacillus bulgaricus* to milk to produce lactic acid and thereby yogurt. Or using genetically modified bacteria to make insulin or human growth hormone to treat diabetes and dwarfism. In the case of aspartame, *Bacillus thermoproteolyticus* is used to join together the two amino acids that make up aspartame. I suspect the question was asked because of concern that bacterial waste implies some sort of safety issue. It does not. The safety of a product does not depend on the route used to produce it; it is established by extensively studying it in the laboratory, in animal models, and by monitoring its use in humans.

Then there was a question about whether ripe bananas, full of black spots, have anticancer properties. The answer is simple: no. The daft banana story was based on a Japanese study reported in an obscure journal that involved injecting banana extract into the peritoneal cavity of rodents. Why this was done isn't clear. Probably because nobody had done it before. Scientists will be scientists. So what did they find? A slight increase in the animals' production of tumor necrosis factor (TNF), which as the name suggests, can have antitumor effects. But it can also have negative effects, such as exacerbating arthritis. In any case, this rodent study has no relevance to humans. We eat bananas; we don't mainline them. Furthermore, the circulating email suggests that bananas contain TNF, which is nonsense. Even if they did, it would not matter because this is a protein that would be broken down during digestion. Bananas make for a great snack, but there is no point in looking for spotted ones with the hopes of preventing cancer.

Now for the really weird. A worried woman asked if it is safe to drink milk when she travels to the U.S. Why? She had heard that women who drink milk from cows treated with bovine somatotropin to increase milk production are at risk for growing mustaches. Apparently, that story came from some Russian official who claimed that American milk causes women to develop male sexual characteristics. What a load of twaddle. Bovine somatotropin is not bioactive in humans. The only mustache you'll get by drinking milk is made of milk.

Next question. How do I know my answers are reasonable? Because the Internet allows me to search virtually all the published literature without leaving my desk. Clearly, the era of the printed encyclopedia is over. Although I was emotionally attached to my copy, it was taking up so much space that I decided to give it away. Nobody wanted it. A sign of the times.

WHAT'S FOR DINNER?

So, what should we have for dinner? It seems a simple question. But it is ever difficult to answer! Unfortunately, tasty and healthy don't always coincide. And just what is "healthy," anyway?

For close to forty years, I've pored through countless research papers and media accounts about food and nutrition. I've interviewed some of the world's top researchers in this area. My shelves sag with the weight of dozens and dozens of books on the subject, and this is the second book I've written myself that deals exclusively with food. My fifteen other books deal with plenty of nutritional connections as well. And even after all that, I'm still mystified about what we should have for dinner. But not completely. The wheat is slowly being separated from the chaff.

One thing is for sure: there's no shortage of nutritional information or opinions about what we should eat. Dr. Robert Lustig, a pediatric neuroendocrinologist at the University of California, believes that many health problems, obesity in particular, can be traced to consuming too much fructose. His video on the subject has gone "viral." University of Missouri professor Frederick vom Saal is of the opinion that obesity can be linked to bisphenol A, a chemical that can leach from the lining of canned foods. Dermatologist Dr. Robert Bibb, in his book *Deadly Dairy Deception*, makes a case for dairy products being the cause of prostate and breast cancer. Dr. Neal Barnard, president of the Physicians Committee for Responsible Medicine, goes even further in *Eat Right, Live Longer*, claiming that salvation lies in avoiding all animal products.

Cardiologist Dr. William Davis sees no problem with meat but sees wheat as the real bogeyman. According to him, the grain's polypeptides cross the blood–brain barrier and interact with opiate receptors to induce a mild euphoria that in turn causes addiction to wheat. As he describes in his bestselling book *Wheat Belly*, this results in fluctuating blood sugar levels that then create hunger and lead to obesity as well as numerous other health problems. How does Davis know all this? Apparently, his patients lose weight on a wheat-free diet and recover from all sorts of diseases. Has he published any of this in peer-reviewed journals? Not that I can find.

Davis would probably find a kindred spirit in Dr. Drew Ramsey, a clinical professor of psychiatry at Columbia University in New York. His book is called *The Happiness Diet: A Nutritional Prescription for a Sharp Brain, Balanced Mood, and Lean, Energized Body*. What is that wondrous prescription? It seems simple enough. If you want to be happy, stay away from bagels. According to Dr. Ramsey, "At first bagels boost a person's energy, but after a few hours you come crashing down looking for another fix in the modern American diet. That crash can cause people to feel irritable, lightheaded, or sad." Really? Maybe if they eat American bagels. I think legions of happy Montreal bagel lovers would disagree.

Journalist Gary Taubes maintains that not only wheat, but all carbohydrates, should be limited. In *Good Calories, Bad Calories*, he has gathered a massive amount of information to "prove" that excessive consumption of carbohydrates is the cause of heart disease, cancer, Alzheimer's disease, and type 2 diabetes. He advises against a low-fat diet. Dr. Dean Ornish, in *Eat More, Weigh Less*, would take issue with Taubes. He puts his cardiac patients on an extremely low-fat, high-complex-carbohydrate diet and has evidence that deposits in arteries actually regress.

I could go on and on about all the dietary advice that floods us. In *Soy Smart Health*, Dr. Neil Solomon claims that eating soy can decrease the risk of breast cancer, heart disease, and osteoporosis, while in *The Whole Soy Story*, Dr. Kaayla Daniel links soy to malnutrition, digestive problems, thyroid dysfunction, cognitive decline, reproductive disorders, heart disease, and cancer. Go figure.

Other authors suggest that our health is being undermined by monosodium glutamate (MSG) or artificial sweeteners or trans fats or pesticide residues or cooking in Teflon pans or genetically modified organisms or chlorinated water or acrylamide or phthalates or hormone residues or . . . or . . . or . . . And there is no lack of advice about how to get our health back on track. All we have to do is drink some esoteric juice, pop some sort of dietary supplement, gorge on some superfood, or eat like the Greeks or Chinese. So who do we listen to? The "experts" can't all be right.

When I throw all the divergent opinions into my mental flask and distill the essence, I come up with something like *The Okinawa Diet Plan*. This fascinating and very well-researched book chronicles the lifestyle habits of the longest-lived population in the world. We're not looking at some mythical Shangri-La here. In the Japanese islands of Okinawa, we have our sights on a people whose unusual longevity and good health is well documented. So is the fact that Okinawans do not gain significant weight as they age! Why? Because they consume 1,600 calories a day, at least 500 less than we do. And they do this while eating half a pound more food. It's all a matter of what sort of food: no hamburgers, hot dogs, or smoked meat here. And no soda pop. But they do eat plenty of food with very few calories per gram.

The lower the calorie density, the more food can be eaten without gaining weight. Basically, this means a plant-based diet. For example, broccoli, mushrooms, and carrots check in at about 0.4 calories per gram, tofu at 0.7, bread or meat, of which Okinawans eat very little, at about 3.0, and oils weigh in at 8.8. Michael Pollan, in his popular book *The Omnivore's Dilemma*, echoes the Okinawan way of eating: "Eat food. Not too much. Mostly plants." Dr. Shawn Baker, author of the *Carnivore Diet*, would not agree. Oh well . . .

CHRONO NUTRITION

"Eat breakfast yourself, share lunch with a friend, and give dinner away to the enemy" emerged as a proverb in the sixteenth century. That was quite a change from the warning that physicians had been giving since the Middle Ages about breakfast being detrimental to health. You would think that by now, with all the research that has been done, we would have figured out how calories should be distributed throughout the day. Not so.

In 2016, a study comparing the timing of major meals in different countries hit the headlines. In Guatemala and Poland, the largest meal is at lunch, with breakfast and dinner having equal calories. In France, Switzerland, and Italy, breakfasts are small, suppers somewhat larger, but the biggest meal is at lunch. Swedes eat small lunches and their breakfasts and suppers have the same calorie content. Germans, Americans, Danes, Dutch, Belgians, and Canadians eat the largest meal at night. The conclusion that researchers distilled out of these observations was that large evening meals are linked with obesity. But there are caveats. There was insufficient data on snacking, which may contribute to weight gain. Also, it may be that people who eat smaller suppers are more active, perhaps hitting the gym in the evening.

The benefits of eating the largest meal of the day in the morning are supported by a 2017 study that involved over 50,000 Seventh-Day Adventists. They were asked to fill out questionnaires every two years about their dietary habits, physical activity, meal frequency, major health events and changes in body weight. Subjects who ate the largest meal early in the day tended to have a lower body mass index than those whose largest meals were lunch or dinner. Perhaps the most interesting finding was that the lowest body mass index was found in people who ate breakfast and lunch but then did not eat again until the next morning, fasting for some eighteen to nineteen hours.

Extrapolating these findings to the general population is difficult because Seventh-Day Adventists have quite a different lifestyle from the average person. They consume no alcohol, eat less meat, and many are vegetarian. Still, the fasting aspect is interesting. With a steady supply of carbohydrates being cut off, the body undergoes a metabolic shift and starts using fat as fuel instead of carbohydrates, leading to fat loss.

The benefit of a large breakfast and small supper gets a boost from a study in Israel that put overweight women on an unevenly distributed 1,400-calorie-a-day diet. Half the women consumed 700 calories at breakfast, 500 at lunch, and 200 at supper, with the other half following the diet in reverse order.

Both groups lost weight, but those who ate the large breakfast lost two and a half times more weight and lost more belly fat. Furthermore, their fasting glucose levels improved. It seems that when it comes to weight loss, when you eat may be as important as what you eat! At least according to this study.

A mouse study further supports the importance of "chrono nutrition," the idea that timing of meals has an effect on physiology. When mice are given unlimited access to a high-fat diet they become obese in about ten weeks and develop high blood cholesterol and insulin resistance. However, when they have access to the same diet for only eight hours a day, they do not become obese or diabetic even though they consume the same number of calories as the mice that have free access to food.

More evidence about the benefits of restricting calories according to a certain time frame comes from a study in which fifty people were told that they could eat anything they wanted for a month except for five days when they would be allowed only about 1,000 calories. After three months, fasting blood sugar, cholesterol, triglycerides, abdominal fat, and markers for some cancers improved significantly when compared with a control group.

Intriguing results have also been found for the 5:2 diet, in which people eat whatever they want for five days and then for two consecutive days limit their diet to 500 calories. There is better glucose regulation and greater loss of belly fat when compared with people who consume the same calories evenly distributed through the week. On the other hand, Courtney Peterson of the University of Alabama found that there was nothing special about squeezing all your daily meals into a brief time slot. She had one group of men eat all their meals within a six-hour period, while a control group consumed the same number of calories over twelve hours. If there were something physiologically special associated with fasting, men in the experimental group should have lost weight. They did not. Peterson's conclusion was that people who lose weight on an "intermittent fasting" diet do so because they end up eating less.

But wouldn't you know it, just when you think you are getting a leg up on all this, you hear about claims that eating whatever you want as long as you eat only between 5 p.m. and bedtime, another example of intermittent fasting, leads to weight loss and greater vigor! And on top of that, a meta-analysis of trials that have examined the effect of eating breakfast on weight change concludes that there is insufficient evidence for recommending the eating of a big breakfast for weight loss.

What about skipping breakfast altogether, as many people do? Not a good idea, at least if we go by a 2019 study published in the *Journal of the American College of Cardiology* that reported on a survey in which between 1988 and 1994 some 6,550 adults were asked if they consumed breakfast every day, some days, rarely, or never. By 2011, 619 of the people who completed the original survey had died from stroke or heart disease. After controlling for age, smoking, alcohol consumption, obesity, socioeconomic status, and activity level, the researchers determined that there was a significantly greater risk of stroke among subjects who never ate breakfast than those who ate it everyday. People who ate breakfast some days or rarely did not have an increased risk of stroke.

A study like this can only demonstrate an association between skipping breakfast and risk of stroke, not a cause and effect relationship. There are other considerations worth noting as well. The subjects were asked only once at the beginning about their breakfast habits and it was assumed that these habits were maintained over the follow-up period, which is questionable. Furthermore, the subjects were not asked about what they ate. One would assume that a sugary pastry gulped down with coffee does not have the same effect on health as oat bran topped with a generous serving of blueberries. Headlines such as "Skipping Breakfast Tied to Early Death" do not properly reflect the nuanced findings of this study,

My takeaway, for whatever it is worth, is that it is better not to have the biggest meal of the day in the evening and having breakfast is advisable. I go back to another sixteenth-century proverb: "Eat breakfast like a king, lunch like a prince, and dinner like a pauper."

ADVICE ABOUT FOOD IS SOMETIMES HALF-BAKED

Back in the early 1970s, just as I was developing an interest in the chemistry of food, I came across a witty quote by Mark Twain. "Part of the secret of success in life is to eat what you like and let the food fight it out inside." Twain was likely reacting to the plethora of health fads that were rippling through America at the time. As evidenced by a passage in his classic work *The Adventures of Tom Sawyer*, he didn't approve: "[Aunt Polly] was a subscriber for all the 'Health' periodicals and phrenological frauds; and the solemn ignorance they were inflated with was breath to her nostrils. All the 'rot' they contained about . . . what to eat, and what to drink, and how much exercise to take, and what frame of mind to keep one's self in . . . was all gospel to her, and she never observed that her health-journals of the current month customarily upset everything they had recommended the month before."

Indeed, there was health advice galore in the nineteenth century. Sylvester Graham urged people to eschew white flour, cooked vegetables, and meat. Drinking water during a meal was verboten. If a vegetarian and a meat eater were shot and killed, Graham maintained, the body of the vegetable eater would take two to three times as long to become intolerably offensive from the process of putrefaction. There is no record of Graham ever testing this theory. Dr. John Harvey Kellogg followed in Graham's footsteps, curing the rich and famous of diseases they never had with a regimen of vegetables, fruits, over-baked bread, and yogurt.

Horace Fletcher, the "Apostle of Correct Nutrition," suggested that the secret to good health lay in chewing food until the last hint of flavor disappeared, and Lydia Pinkham promoted her Vegetable Compound as just the thing for "female complaints and weaknesses." Dr. James Salisbury claimed that heart disease, tumors, mental illness, and tuberculosis were the result of vegetables and starchy foods producing poisonous substances in the digestive system. His solution was the Salisbury steak, essentially fried ground beef with onion and seasonings. According to the good doctor, the steak was to be eaten three times a day with lots of water. This would cleanse the digestive system and, as a bonus, the high-meat, low-carbohydrate diet would lead to weight loss. Early shades of Atkins.

Little wonder that Mark Twain poked fun at these half-baked, contradictory fragments of advice with his suggestion to let the food fight it out once inside. That of course was pure whimsy, but foods really do duke it out, though not inside our bodies. Rather, it is in the scientific literature that dietary components vie for infamy or honor. And the biggest battles take place when the stakes are high, such as in the struggle against heart disease.

I've now been watching that battlefield for more than four decades. I have imersed myself in books about the relationship between diet and heart disease, ranging from *The China*

Study, in which Dr. T. Colin Campbell urges us to reduce blood cholesterol by eliminating all animal products, to Dr. Malcolm Kendrick's *The Great Cholesterol Con* and Dr. Ernest N. Curtis's *The Cholesterol Delusion*, which claim that a high-fat diet does not put a person at risk for coronary artery disease and that lowering the cholesterol level with diet or drugs will not prevent heart attacks. My filing cabinets swell with the studies referenced in these books plus numerous others. One would think that a definitive conclusion about the relationship between diet and heart disease could be arrived at by digging through all this material. Alas, it is possible to find reputable studies to either support or oppose the obsession with cholesterol. When it comes to dueling studies, there is rarely a clear-cut winner.

When I began my search for light at the end of the misty tunnel of nutrition oh so many years ago, one name kept cropping up. Ancel Keys was a physiologist who had noted that well-fed American businessmen suffered a higher rate of heart disease than postwar undernourished Europeans. Keys knew that atherosclerosis was characterized by deposits of cholesterol in the walls of the arteries, and that in the early 1900s Russian scientist Nikolay Anichkov had shown a link between feeding cholesterol to rabbits and artery damage. He was also aware that in the 1940s John Gofman had identified lipoproteins as the molecules that transport cholesterol through the bloodstream and that he had demonstrated a relationship between blood levels of these lipoproteins and the risk of heart disease.

Since cholesterol is present in the human diet, mostly in fatty animal foods, Keys thought a relationship between diet and heart disease was likely. One way to explore this possibility was to compare disease patterns in countries with different amounts of fat in the diet. In his famous Seven Countries Study, Keys showed that both elevated mean blood cholesterol levels and deaths from heart disease correlated with the percent of calories attributed to fat in the diet, although there were a few exceptions. Inhabitants of the island of Crete had the lowest heart disease rate but ate lots of fat. Their fat intake, however, was mostly of the unsaturated variety found in fish and olive oil. So Keys concluded the real culprit was saturated fat and promoted a Mediterranean diet, emphasizing unsaturated over saturated fats.

Critics quickly pointed out that increased heart disease rates correlated even better with the number of radios produced or with the amount of gasoline sold, highlighting that an association cannot prove that a cause and effect relationship exists. There were also questions about the reliability of death certificates to determine heart disease mortality, as well as about the calculation of fat consumption. Then there was the bothersome point of Keys choosing only seven countries when statistics about food consumption and mortality were available for at least twenty-two others. Did he leave these out because the data did not fit the straight-line relationship that was evident when only seven countries were considered? And with that salvo of criticisms, the war between the pro- and anti-fat forces was launched.

Keys correlated the risk of death from heart disease with levels of blood cholesterol and with the amount of saturated fat in the diet. This did beg for further exploration. That was undertaken with the famous Framingham study that followed more than 5,000 initially healthy inhabitants of the small Massachusetts town and confirmed that high blood cholesterol correlated, albeit weakly, with heart disease. Rarely mentioned, however, is the fact that the Framingham study found no relation between fat consumption and heart disease! Observational studies such as Keys's and Framingham can only show associations. To prove that high cholesterol is a causative factor in heart disease, and that it is a function of diet, an intervention study is needed. A demonstration that a low-fat, low-cholesterol diet results in a drop in blood cholesterol and also parallels a decline in heart disease would constitute good evidence for recommending such a diet. In 1972, the Multiple Risk Factor Intervention Trial, cleverly abbreviated as MRFIT, took on this challenge. Some 12,000 men at high risk for heart disease because of high cholesterol, elevated blood pressure, and smoking habits were divided into two groups. One group got advice on quitting smoking, management of high blood pressure, and received intensive instruction on preparing food that was low in cholesterol, low in saturated fat, and high in polyunsaturated fat. The other group received no specific advice other than what would normally be offered by their family physician.

After ten years, the intervention group had reduced saturated fat intake by almost 30 percent and increased polyunsaturated fats by 33 percent, while the diet of the control group was essentially unchanged. Blood pressure was reduced significantly in the intervention group and about half the smokers gave up the habit. But in spite of the intense changes in diet, total cholesterol declined by only 7 percent. At the end of the study, there were 115 deaths ascribed to heart disease in the intervention group and 124 in the control group. Although that was significant, the result was clouded by the fact that there were 265 total deaths in the intervention group as opposed to 260 in the control group. Rigorous modification of risk factors had not provided the impactive results that had been hoped for.

Roughly at the same time as MRFIT, the Lipid Research Clinics Coronary Primary Prevention Trial (CPPT) enlisted some 3,800 men with cholesterol levels that ranked in the top 1 percent of the population. Half were given cholestyramine, a drug expected to reduce cholesterol significantly. After ten years, the actual reduction was only 8 percent, but this did result in a 19 percent reduction in non-fatal heart attacks and coronary disease deaths. The authors' claim that each 1 percent reduction in cholesterol would result in a 2 percent reduction in cardiac risk made for splashy headlines.

But where exactly does 19 percent come from? In the treatment group, 7 percent of the subjects died or had a heart attack, while the corresponding figure in the control group was 8.6 percent. Decrease 8.6 by 19 percent and you get 7 percent! Another way of saying this is that one would have to aggressively treat high cholesterol in about sixty-seven people to save one cardiac event. Not very impressive, but still, Keys's correlation, together with CPPT and MRFIT, were judged to have provided enough evidence to justify making recommendations to the public about reducing blood cholesterol to reduce the risk of heart disease. That advice centered on manipulating the amount and type of fat in the diet. Incredibly, after some fifty years of research, it still is not clear just what that manipulation should be. Well, perhaps it isn't so incredible. It is very difficult to tease meaningful results out of epidemiology, especially when it comes to nutrition. Statistics about disease patterns are often ambiguous and people's recall of what and how much they ate is notoriously unreliable.

On one issue, there is agreement. Trans fats are a risk for heart disease and eliminating them amounts to good riddance. But what about saturated fats? Here the issue is not so clear. In 2010, the *American Journal of Clinical Nutrition* featured an analysis of twenty-one major studies and concluded that there was no significant evidence associating saturated fat with an increased risk of heart disease. A Japanese study in the same journal surprisingly found that as saturated fat intake increased, the risk of stroke actually decreased! On the other hand, Harvard researchers examined eight randomized controlled trials in which saturated fats were replaced with polyunsaturated fats and found a modest protection against heart disease. But even such a replacement is not so straightforward, as outlined in a 2013 paper published in the *Canadian Medical Association Journal*. Oils like corn or safflower oil, which are rich in omega-6 fats but poor in omega-3 fats, should not be promoted as reducing the risk of heart disease, while making such a claim for oils such as canola or soya, which are rich in both these fats, is reasonable.

Although the evidence for reducing the risk of heart disease by manipulating the fat content of the diet is less compelling than is generally assumed, society's fear of fats has resulted in numerous non-fat and low-fat products in the marketplace. This in spite of the fact there is no good data to show that people diagnosed with coronary disease have consumed more fat than healthy people and that more than half of all heart attack victims have normal or low blood cholesterol. Given that in low-fat foods the fat ends up being replaced by various carbohydrates, often simple sugar, we may have gone from the frying pan into the fire.

That's just what Dr. John Yudkin, a contemporary and critic of Ancel Keys, suggested. In his book *Sweet and Dangerous* (originally published in the U.K. as *Pure, White and Deadly*), adorned with a sugar bowl sporting a skull and crossbones, Yudkin pointed out that the correlation between sugar consumption and heart disease was stronger than the one between fat and heart disease. His view was almost universally dismissed but now is being resuscitated with further evidence from Dr. Robert Lustig, who links sugar not only to heart disease but to obesity as well. Having followed the "cholesterol hypothesis" for more than four decades, I still can't come to a firm conclusion, but evidence is mounting that sugar is a greater villain than saturated fat. As Mark Twain said, "It ain't what you don't know that gets you into trouble. It's what you know for sure that just ain't so."

DINING ON LIQUID GOLD

The Greek poet Homer was in all likelihood not referring to health benefits when he called olive oil "liquid gold." The Mediterranean diet, with its health implications and emphasis on olive oil, would not hit the headlines for another 3,000 years.

Keys's findings in the Seven Countries Study were enough to convince the American Heart Association to promote a diet that aimed to reduce the risk of heart disease by reducing fat intake. To cut down the risk of heart disease, the message went, cut down on butter, lard, eggs, and beef. But the fact that saturated fat intake was not the only difference between the traditional American diet and the eating habits of the Greeks and Italians was not addressed. It took the Lyon Diet Heart Study in 1994 to demonstrate that the American Heart Association diet and a Mediterranean diet had quite different clinical outcomes.

Researchers in France investigated patients who had a heart attack and were subsequently counseled to follow a Mediterranean diet. They were encouraged to eat more fruits, vegetables, and fish, reduce their red meat consumption, and to replace butter with a special margarine that was formulated to contain the type of fats found in the Cretan diet. Why margarine? Because the researchers thought that the French palate, accustomed to eating butter, might accept margarine but not olive oil. Also, the margarine contained alpha-linolenic acid, an omega-3 fat that is prominent in the Cretan diet with its emphasis on walnuts, olive oil, and a vegetable called purslane. After just two years, the death rate in the intervention group was reduced by 70 percent! This approach, which controlled the type but not the amount of fat consumed, was clearly superior to the Heart Association's low-fat diet. Surprisingly, there was no difference in the blood cholesterol levels of the groups.

Then in 2003 came a prospective investigation of some 22,000 Greek adults by University of Athens researchers who developed a point system to assess how closely the subjects followed the traditional Mediterranean diet. Fruits and nuts, for example, were awarded points, and meat, poultry, and high-fat dairy products were not. After nearly four years, 275 of the subjects had died, most of whom had not been following the Mediterranean diet. The more closely subjects followed the diet, the less likely they were to die during the four-year trial.

While no single food was predictive of mortality, increased intake of fruits and nuts was associated with greater chance of survival, as was an increased ratio of monounsaturated to saturated fats. Such an increased ratio is a reflection of a greater intake of olive oil and a reduced consumption of meat. Unfortunately, any study that relies on self-reported food intake is burdened by recall bias. In this case, the subjects were asked to report the frequency of their intake of 150 foods and beverages during the year preceding the study. With such studies, there is always a problem with inaccurate recall. Furthermore, it is quite possible that some of the subjects changed their diet significantly over the four-year period.

A recent randomized intervention trial published in the *New England Journal of Medicine* addressed these concerns. Spanish researchers followed 7,447 men and women who were free of heart disease but had either type 2 diabetes or at least three risk factors defined as smoking, hypertension, elevated LDL ("bad cholesterol"), low HDL ("good cholesterol"), being overweight, or having a family history of coronary heart disease. Three diets were randomly assigned to the subjects: a Mediterranean diet supplemented with extra-virgin olive oil, a Mediterranean diet supplemented with mixed nuts, or a control diet that just aimed to reduce fat in general.

The trial was stopped after 4.8 years, earlier than planned, because the experimental groups were showing a statistically significant reduction in cardiovascular events. Interestingly, there was very little difference in fat consumption between the groups. According to the researchers, the benefits were derived from the 50 grams (4 tablespoons) of extra-virgin olive oil a day and the six servings of nuts a week. (A serving consisted of 15 grams of walnuts, 7.5 grams almonds, and 7.5 grams hazelnuts.) There was also a slight increase in fish and legume intake in the experimental groups.

What's the takeaway message for North Americans? It seems that it is the type of fat rather than the amount that is important. This notion is supported by the theory that inflammation plays a more important role in heart disease than blood cholesterol. Trans fats that lurk in many processed foods and omega-6 fats, as found in corn, soybean, and sunflower oil, all popular in our diet, are inflammatory, while nut oils and olive oil have an anti-inflammatory effect.

The anti-inflammatory effect of olive oil may be double-pronged. Monounsaturated fats appear to curb inflammation, but a compound called oleocanthal isolated from extra-virgin olive oil has been shown to have the same effects as the anti-inflammatory medication ibuprofen. This discovery came about in a fascinating way. Gary Beauchamp of the Monell Chemical Senses Center in Philadelphia was attending a scientific meeting in Italy, when he tried some freshly pressed extra-virgin olive oil. The experience wasn't a pleasurable one, as almost immediately he began to feel a stinging sensation in his throat. As luck would have it, Beauchamp had previously worked on testing the sensory properties of ibuprofen medications and had experienced exactly the same effect. This made Beauchamp wonder: Could there be some connection between olive oil and ibuprofen?

Oleocanthal was eventually isolated from olive oil and was determined to be responsible for the stinging effect. Curiously though, this compound had no chemical resemblance to ibuprofen. There seemed no doubt though that oleocanthal was the stinger, since a synthetic version added to corn oil also resulted in throat irritation. Further experiments revealed that oleocanthal blocked the action of the enzymes known as COX-1 and COX-2, which are known to produce inflammation. The next question to answer was just how much oleocanthal there is in olive oil and how much of the oil would have to be consumed for an effective dose. The answer turned out to be a great deal! A whole glass of olive oil would be needed to treat a headache. This of course is not recommended but incorporating some extra-virgin olive oil into the diet is a good idea, especially given that the oil also contains a variety of polyphenols with antioxidant properties. This may be why some epidemiological studies have suggested that people who have a diet rich in extra-virgin olive oil may have a lower risk of breast and colon cancer.

And let's not forget that sugar, which has also been linked to inflammation, is scarce in the Mediterranean diet. So extra-virgin olive oil, nuts, and low sugar and meat consumption may be the keys to cardiovascular health. But just how much protection can people with risk factors expect if they make dietary changes as in the Spanish trial? Surprisingly little. Roughly 100 people have to change their diet to prevent one cardiovascular event. Of course that is pretty significant if you are the one.

The bottom line, then, is that it pays to eat like the Cretans and other Mediterraneans. But the North American version of the Mediterranean diet will not do. Fried calamari most assuredly will not make you live longer.

SOME BEEFS WITH BEEF

These days a lot of people have a beef with beef. There is the ethical issue of raising animals for slaughter, an issue that cannot be settled by science. But science can certainly provide insight into the environmental impact of raising livestock, as well as into questions about the effects of eating beef on our health.

Raising cattle is not an environmentally friendly process, about that there can be no debate. First, there is the matter of cows burping methane, a greenhouse gas, then there is the amount of water needed to raise cattle. Every serving of beef requires about 2,000 liters of water. Skip a burger, and you can save enough water to shower with for at least a month.

While cattle usually start out grazing on grass or eating silage, they are "finished" on feedlots with corn or soy, grains that can also feed people. About 45 percent of all grain grown goes towards feeding animals. The oft-quoted statistic is that for cattle raised on feedlots, as is commonly the case in America, about 7 kilograms of grain are needed to produce 1 kilogram of meat. Obviously, cows reared on grass and hay do not take food out of human mouths, but they do emit about twice as much methane as cattle raised on a feedlot since they grow more slowly and live longer. The world has a lot of pastureland where crops can't be grown, and calculations show that feeding livestock residues from other crops coupled with sustainable grazing could still provide about two-thirds of the meat that is currently produced. But given the world's growing appetite for meat, this is not likely to happen.

On the plus side, beef is an efficient way of providing protein relative to total calories. An adult requires roughly fifty grams of protein a day, an amount that can be provided by 200 grams of steak (400 calories). It would take 600 grams of kidney beans (760 calories) to provide the same amount of protein. Meat is also a good source of B vitamins, zinc, and iron. But there is an issue with the iron found in meat.

Since the 1970s, researchers have noted a greater incidence of colorectal cancer in countries where beef consumption is high. The same does not hold for poultry or fish. What is the difference? One theory is that it is the iron content of beef, twice that of poultry, that may be responsible. Iron in meat is bound to heme, the non-protein part of hemoglobin, the compound that carries oxygen around the bloodstream. Heme iron is absorbed more readily than iron in plant products, and laboratory studies show that this form of iron oxidizes fats to yield peroxyl radicals, compounds that can effectively cleave DNA, a possible initial step in triggering colon cancer.

Not all studies show an association between colorectal cancer and fresh meat, but virtually all show that link with processed meats. That may have to do with the nitrites used as preservatives being converted into carcinogenic nitrosamines. In 2013, data from the European Prospective Investigation into Cancer and Nutrition study (EPIC) revealed that for every 50 grams of processed meat consumed a day, the risk of early death from all causes during the twelve-year period of the study increased by 18 percent. Researchers had followed half a million people, distinguishing between consumption of red meat, white meat, and processed meat, while controlling for factors such as smoking, weight, fitness, and education levels, all of which can influence health. When it came to fresh, unprocessed meat, there was no association with ill health. Indeed, people who ate no meat at all had a greater risk of early death than people who ate a little meat! Furthermore, the risk of early death was significantly reduced among meat eaters who ate a lot of fiber. In the U.S. the National Health and Nutrition Examination Survey (NHANES) has followed over 18,000 people for decades and has found no link at all between meat consumption and ill health.

The way meat is cooked can raise concerns. High temperatures produce heterocyclic aromatic amines, polycyclic hydrocarbons, and advanced glycation end products. Suffice it to say that these are nasty compounds, concentrated in charred parts of meat, that we can do without. Lower cooking temperatures are desirable, but then the problem of bacterial contamination rears its ugly head, particularly with burgers since grinding can spread bacteria that normally are found only on the surface of meat throughout. Animals, like humans, harbor a host of bacteria, most of which are harmless. But not all. *E. coli* O157:H7, naturally found in the intestinal contents of some cattle, goats, and even sheep, can cause a condition known as hemolytic uremic syndrome (HUS) in which red blood cells are destroyed and kidney failure ensues. Back in 1993, four children died and 178 other people suffered permanent kidney or brain damage after consuming undercooked hamburgers at Jack in the Box restaurants in the U.S. This prompted an outcry for hamburgers to be cooked to an internal temperature of at least 70 degrees Celsius.

What do we distill out of all this? Cutting back on eating meat is beneficial for health as well as for the environment. According to an estimate by the World Economic Forum if everyone in the world gave up beef there would be a 25 percent reduction in global greenhouse gas emissions, and diet-related deaths would drop by 5 percent. Seems it is better to beef up our protein intake with beans, nuts, algae, and perhaps even insects.

COUNT YOUR WAY TO GOOD HEALTH

Would you like to reduce your risk of cancer? You can, if you can count to five. If you can make it to ten, you'll reduce your risk even more. The secret is not in the mental gymnastics; it's in what you will be counting. I am constantly amazed by people who swallow a surfeit of unproven dietary supplements, hoping to protect themselves from cancer. Instead, they could be following a relatively simple regimen that has been scientifically shown to reduce risk by up to 50 percent. I am talking, of course, about consuming five to ten servings of fruits and vegetables every day.

Scientists, as you know, waffle on many issues. We're not sure about the risks associated with radiation from cell phones, mercury in fish, or fluoride in water. We debate the extent to which greenhouse gases need to be controlled. We puzzle over the use of estrogen supplements and often don't agree on the pros and cons of genetically modified foods or pesticides. This vacillation may seem surprising, because the scientific literature on each of these issues is abundant. But these are complex topics, with enough "on the one hand" and "on the other hand" arguments to preclude categorical conclusions. It is rare indeed to find a subject on which there is virtually universal scientific agreement. Rare, but not impossible! One would be hard-pressed to find a scientist who opposes increased consumption of fruits and vegetables. Why? Because the evidence is as ironclad as is possible in the often-confusing world of human nutrition.

What constitutes such strong evidence? Hundreds and hundreds of studies carried out around the world have furnished us with data demonstrating that fruit and vegetable intake reduces the risk of cancer. An analysis of these studies proves the point and also allows us to learn about the various types of studies that scientists design in order to be able to offer meaningful nutritional advice.

Information about a potential link between diet and disease usually first emerges from what is called a cohort study. Detailed food frequency and health status questionnaires are mailed to thousands of healthy people periodically over many years; the responses are then analyzed to see if nutritional patterns can be linked with specific diseases. A classic example is the Nurses' Health Study, which began in 1976 when 121,700 female nurses aged thirty to thirty-five started to fill out biennial questionnaires. After just ten years, it became apparent that subjects who consumed the most carotenoid-containing fruits and vegetables had a 20 to 25 percent lower risk of lung cancer than those who ate smaller amounts. Carotenoids are familiar as the yellow and red pigments found in carrots and tomatoes. In another cohort study, over 1,200 Massachusetts residents over the age of sixty-five were asked to report on their diets in 1985 and were then followed by researchers. Those who eventually died from cancer ate the fewest vegetables and those who ate the most green and yellow vegetables had the lowest cancer rates.

The most plentiful information about the link between diet and cancer comes from "case-control studies." Persons with a particular disease are identified and asked about their past dietary history and other possible risk factors. They are then matched against a similar group of control subjects, those who do not have the disease. In a typical case-control study, 179 pancreatic cancer cases were matched with 239 controls. Patients with the disease were more likely to have consumed more smoked and fried foods and far less raw fruits and vegetables. Over 150 such studies have shown that fruits and vegetables provide a protective effect against various cancers. The greatest evidence of this effect is for cancers of the stomach (the most common cancer in the world), lung, and esophagus.

Animal-feeding studies and test-tube experiments also offer insight. The animals in question are usually rats or mice that are fed a diet of certain fruits or vegetables before they are exposed to a chemical carcinogen. Their tendency to develop tumors is then investigated. Perhaps the most alluring example here comes from the work of Dr. Paul Talalay, director of the Brassica Chemoprotection Laboratory at Johns Hopkins University in Baltimore, Maryland. Stimulated by a number of cohort studies that showed an inverse correlation between cancer and the consumption of cruciferous vegetables, such as cauliflower and broccoli, Dr. Talalay sought to isolate the protective factor. His research pointed at sulforaphane, a compound that forms from its precursor, glucoraphanin, when broccoli is chewed. He went on to feed sulforaphane to rats that had been exposed to dimethylbenzanthracene, a potent carcinogen found in smoke. Almost 70 percent of the control rats developed cancer, but only 35 percent of the animals that had dined on the broccoli extract did. Dr. Talalay then added sulforaphane to cultured human cells in the laboratory and again demonstrated protection against cancer.

We are even beginning to see some human "intervention studies" using vegetables. These investigations are potentially the most compelling. Because of the reputed protective effect of tomato products against prostate cancer, fifteen patients at

Wayne State University in Michigan were given 15 milligrams of lycopene, the red pigment in tomatoes, twice daily for three weeks prior to surgery. The patients' tumors were smaller and showed reduced malignancy compared with the tumors of patients who had been given a placebo.

While no single study can, or should, be convincing about the protective effect of fruits and vegetables, there is no doubt that the preponderance of evidence is overwhelming. Still, most North Americans consume three or fewer servings per day. If we doubled our intake of fruits and vegetables instead of searching for solace in the latest multilevel-marketed "miracle," we would likely see a drop in cancer incidence.

So all together now, let's count: (1) half a cup of green beans, (2) a couple of carrots, (3) half a cup of broccoli, (4) half a cup of grapes, (5) a banana, (6) an apple. That's not so hard to do, is it? And don't forget the tomato sauce. You can add some garlic to it for further protection against cancer — and vampires.

BEATING THE ODDS WITH BEETS

Who doesn't love a Cinderella story? And we certainly had one in 2016 when Leicester City, a team that finished in fourteenth place the previous year, won the English Premier League championship. How did Leicester overcome odds of 5,000 to 1 and rise to the top in one year? According to statistics, during that fairy-tale year, the team had the fewest injuries and used the fewest players in the league. When it came to sprint tests, Leicester City had the quickest players, including Jamie Vardy, who recorded the fastest speed in the Premier League at thirty-five kilometers per hour. Had Leicester found some sort of magical formula for winning?

The team's trainers emphasized lots of forty-meter sprints and exercises to build up hamstring muscles. But all teams do that. However, not all teams follow sprints with a stint in a cryotherapy chamber, exposing players to air cooled to about minus 135°C. Such whole-body cryotherapy constricts blood vessels and reduces blood flow to the extremities, which is claimed to reduce inflammation around soft-tissue injuries since fewer inflammation-causing white blood cells reach these sites. The contraction of the blood vessels also supposedly leaves more blood in the lungs where it is oxygenated. When the body warms, the freshly oxygenated blood is circulated to the muscles, preventing soreness. Furthermore, the stress of the extreme cold is said to result in the release of adrenalin, the fight-or-flight hormone that can relieve pain and cause a feeling of being energized.

When it comes to science, though, claims about the benefits of cryotherapy outstrip the evidence. A Cochrane review, widely regarded as a reliable compilation of evidence-based studies, found that there was insufficient evidence for cryotherapy relieving muscle soreness after exercise. Evidence that the therapy improves athletic performance is also lacking. Same goes for claims about treating arthritis, fibromyalgia, multiple sclerosis, sleep disorders, and depression. But there is agreement that the torturous experience of being immersed in extremely cold air can produce a powerful placebo effect.

If Leicester's miraculous performance is unlikely to have been due to cryotherapy, what other factor could have been involved? It turns out that all season the players were drinking beetroot juice!

There is a surprising amount of research on beetroot juice, with some studies showing that the juice has an anti-inflammatory effect and can reduce the muscle soreness caused by vigorous exercise. Other studies have pointed to beetroot juice possibly increasing stamina. One study featured eight young men who consumed 500 milliliters of beet juice for six days and underwent exercise tests on a stationary bike during the latter three days. The process was then repeated with blackcurrant juice substituting for the beet juice. Oxygen utilization and time to exhaustion were marginally better during the beet juice phase. But surely the subjects could differentiate the taste of the juices they were drinking and could have been aware of the reputed beneficial effects of beet juice. Still, when all studies that used beetroot juice are considered, there seems to be a moderate improvement in exercise endurance with regular consumption.

A potential explanation for the benefits attributed to the juice may be the nitrate content of beets. Nitrates occur naturally in a variety of vegetables such as spinach, celery, cabbage, and beets. Bacteria in the mouth can convert nitrates to nitrites, which are eventually absorbed into the bloodstream, where they serve as a source of nitric oxide. This, in turn, helps to increase blood flow and also allows blood vessels to respond to blood pressure changes with a smaller likelihood of damage. It is interesting to note that the active ingredient in Viagra increases blood flow to the area of interest by boosting nitric oxide production.

Nitrites, it now seems, are good for us. This is surprising after all the adverse publicity about nitrite preservatives in prepared meats! There the argument is that nitrites react in the body to form carcinogenic nitrosamines. Which is it then? Are nitrites good or bad? Maybe, when they're found in fruits and vegetables, we are more likely to see the positive effects because nitrosamine formation is inhibited by the various vitamins and antioxidants present in the vegetables but not in processed meats. Also, proteins in meat are a source of the amines needed for nitrosamine formation, meaning nitrites in meat may be more problematic than nitrites that form in the body as a result of nitrate intake.

Is there a downside to drinking beetroot juice? Well, if you forget you indulged, you may have a scary bathroom experience. The betacyanins responsible for the red color of the beets can make a rather dramatic exit.

What happened to Leicester City after their Cinderella season? One would assume the training methods did not change, but the team dropped back down to twelfth place. So maybe success wasn't due to cryotherapy or beetroot juice. Sometimes in sports, the stars just align.

SWALLOWING BLUEBERRIES, APPLES, AND HYPE

Blueberries may reduce the growth of breast cancer! Apples and pears reduce the chance of stroke! I bet I have your attention now. But those are not my words; they're newspaper headlines. It seems that virtually every day some new study comes out touting the ability of this or that food to extend our earthly existence. Usually, the researchers themselves are modest in their claims and end their discussion with the inevitable call for more research. But then the media get a sniff of the action. And in the drive to capture public attention, science sometimes takes a back seat. Before long, a smidgen of science may be blended with a dash of hope and a healthy dose of hype to cook up a scrumptious headline. But for the scientifically minded, the tasty headline may trigger a bout of mental indigestion.

The blueberry story is a report of an interesting study carried out on female nude mice. Don't get any mental images of Minnie enticing Mickey; these nude mice are specially bred for laboratory research. They derive from a strain with a genetic mutation that causes them to have an underactive thymus gland, resulting in an impaired immune system. Outwardly, they lack body hair, hence the nickname "nude." Suffice it to say that these nude mice are not a perfect model for predicting biological effects in other mice, let alone in humans. Still, they are valuable in research because cancer cells can be introduced without a rejection response. And the blueberry study was all about injecting mice with breast cancer cells. But these were very specific breast cancer cells, known as triple negative cells.

Growth of triple negative cancer cells is not supported by the hormones estrogen or progesterone, and they also test negative for the presence of human epidermal growth factor receptor 2 (HER2), a protein that promotes the growth of cancer cells. Triple negative cells therefore do not respond to standard hormone-blocking drugs such as tamoxifen or to medications such as herceptin, which interfere with the HER2 receptor. Triple negative cells are very aggressive but cause of only 15 percent of all breast cancers.

Now, back to our blueberries. Researchers at City of Hope Hospital in Los Angeles treated nude mice to a diet that included either 5 percent or 10 percent blueberry powder by weight. After two weeks, the mice were injected with the triple negative breast cancer cells. A control group of animals was fed in the same fashion but without the blueberry powder.

Why undertake such an experiment? Because earlier laboratory studies had shown that blueberry extract had anti-angiogenesis activity, meaning that it interfered with the formation of blood vessels that tumors need to grow. After six weeks, the mice fed the 5 percent blueberry diet had a tumor volume that was 75 percent lower than the control animals, but strangely, those fed the higher dose blueberry diet showed only a 60 percent lower tumor volume. In terms of human equivalents, the 5 percent blueberry diet corresponds roughly to eating about two cups of fresh blueberries a day. In a second study, the blueberry-fed mice exhibited a reduced risk of the cancer spreading to other parts of their bodies.

What, then, would be a realistic headline to describe these results? How about "Large Daily Dose of Blueberry Powder May Reduce the Growth of a Rare Type of Artificially Induced Breast Cancer in a Special Variety of Immune-Suppressed Mouse?" That wouldn't sell many papers, one would guess. And what do these mouse experiments mean for humans in terms of preventing or treating breast cancer? Not much. All we can do is mutter that blueberry extracts "warrant further investigation."

Marketers, of course, are not tethered to science. Any blueberry study that hints of some positive outcome, no matter how irrelevant it may be to humans, is enough to trigger an outburst of processed foods that feature blueberries on the packaging. You might think, for example, that Total, a cereal that loudly proclaims "blueberry" on the box, would actually contain blueberries. Well, you would be wrong. The "blueberries" inside are artificially colored and flavored bits of sugar mixed with fat. Even bagels or muffins that actually do have some blueberries contain insignificant amounts for any biological effect. Pure marketing hype.

How about the apple and pear study? Well, it really isn't a study about apples or pears. Researchers at Wageningen University in the Netherlands analyzed food frequency questionnaires filled out by some 20,000 people in terms of their fruit and vegetable consumption. Based upon the color of their "fleshy" portions, the fruits and vegetables were divided into green, orange and yellow, red and purple, and white.

The subjects were followed for ten years, a period during which 233 suffered a stroke. It turns out that stroke victims consumed fewer "white" fruits and vegetables than the other subjects. The researchers calculated that for every 25-gram increase in "white" fruit and vegetable consumption each day, the risk of stroke decreased by 9 percent. This may sound like a significant drop, but really it is a small effect. Out of some 20,000 people, 233 suffered a stroke. That's roughly a 1.2 percent risk. A drop of 9 percent would mean the risk goes down to 1.1 percent. In other words, 1,000 people would have to increase their "white" intake by 25 grams to save one stroke.

So what are "white" fruits and veggies? Bananas, cauliflower, chicory, cucumber, pears, and apples. Within the "white" group, apples and pears were most commonly consumed, hence the catchy headline about apples and pears reducing the risk of strokes.

Now for a splash of critical thinking. First, food frequency questionnaires are notoriously unreliable. People have a hard time remembering what and how much they have eaten. And chances are that dietary habits change over the years. There is no guarantee that the pattern revealed by the questionnaire was followed over the ten-year follow-up period. Next, only white fruit and vegetable consumption was linked to a reduced incidence of stroke, not specifically apple or pear intake. Maybe the effect, if indeed there is one, is due to bananas or cauliflower.

This may sound like we're splitting fruits here. But there is a point. Some reports referred to the 9-percent decrease in strokes for every 25 grams of "white" fruits and vegetables and suggested that eating an apple a day (roughly 120 grams) can reduce the risk of a stroke by some 45 percent. That is some overly exuberant data-dredging; it's akin to inferring that blueberries can reduce the risk of breast cancer based on some mouse experiment.

While both the blueberry and apple studies described are pretty hollow, there are some intervention studies of note.

A study by researchers at King's College in London investigated the effects of blueberry consumption on the health of arteries. A small group of healthy volunteers was asked to consume a daily beverage made with 11 grams of wild blueberry powder, roughly equivalent to 100 grams of fresh wild blueberries. Their blood pressure was regularly monitored, as was the flow-mediated dilation (FMD) of the arteries in their arm. This is a measure of how readily arteries widen as blood flow increases and is a predictor of the risk of heart disease.

After a month there was a significant improvement in FMD as well as a lowering of systolic blood pressure. Similar, although somewhat reduced, effects were found when a mix of pure anthocyanins, the equivalent to the amount in the beverage (160 milligrams), was consumed. It seems that blueberries have some other beneficial components other than anthocyanins as well.

No, this still doesn't make blueberries a "superfood" but at least the study is in step with the plethora of publications that attest to the benefits of eating fruits and vegetables. So by all means fill up on blue, white, and whatever other colored fruits and vegetables you can find. But don't swallow the next headline about some "superfood" saving you from the clutches of the Grim Reaper. Exercise some critical thinking. And get some exercise. That can really reduce the risk of disease.

FIBER AND GRAHAM

Professor John Cowles was not a Grahamite. Back in the 1830s when Oberlin College in Ohio decided to implement and enforce Sylvester Graham's dietary regimen, Cowles expressed his displeasure by smuggling a pepper shaker into the dining room and openly applying its contents to his meal. Soon after this incident the professor was fired, sparking rumors that his fate had been sealed by flaunting the rules. And those rules were stringent. No meat, no sugar, no fat, no alcohol, no spices of any kind. Meals were based on fruits, vegetables, and the hallmark of the Graham system, copious servings of bread made from whole grains.

Presbyterian minister Sylvester Graham can be labeled as America's first nutritional guru, and like the plethora of nutritional advisors who prowl the Internet today, developed a loyal following as well as hordes of critics. Graham had no scientific knowledge, not that there was much of this commodity to be had in the early nineteenth century. His notions were based on the diet that sustained Adam and Eve in Eden, essentially seeds, nuts, berries and fruits. Basically, this was a vegetarian, high-fiber diet, although Graham would have been mystified by the current definition of fiber as the components of plant foods that human digestive enzymes cannot break down, including waxes, lignin, and various polysaccharides. To him it would have been roughage, the key to healthful living, in large part due to its ability to curb sexual excess, particularly in the form of "self-abuse."

Lust, Graham claimed, was responsible for ailments ranging from indigestion to insanity and could be dampened by a diet featuring whole wheat baked goods. Whether Graham actually introduced the cracker named after him is somewhat controversial. His classic book, *A Treatise on Bread and Bread-Making*, published in 1837, discusses the preparation of whole grain bread and condemns the use of additives such as alum to whiten flour but makes no mention of crackers. "Graham crackers" first appeared in the marketplace in the late 1800s, but undoubtedly the "prophet of bran bread" would have approved of the original cracker made of whole wheat.

Sylvester Graham's ideas were embraced by Seventh-Day Adventists and popularized by Dr. John Harvey Kellogg, who headed the Western Health Institute established by the church in Battle Creek, Michigan. Kellogg never consummated his marriage, believing that sex drained the body's health, and was a proponent of treating insanity with a whole grain diet.

While Graham and Kellogg may have been overzealous in their promotion of the health benefits of whole grains, they were on the right track. Kellogg managed to demonstrate the positive effects of wheat bran on patients suffering from constipation and colitis. Then in the 1970s, after studying populations in sub-Saharan Africa, British physicians Denis Burkitt and Hugh Trowell concluded that a lack of unprocessed, high-fiber foods in the typical Western diet leads to a higher incidence of heart disease and colorectal cancer.

The benefits of a high-fiber diet were pushed to the back burner with the emergence of the low-carb regimens such as the keto and paleo diets that effectively targeted extra pounds. However, there may be a price to pay for that weight loss if we consider what should be an impactive study published in 2019 in the *Lancet*, one of the top medical journals in the world.

Researchers led by Professor Jim Mann of the University of Ostego in New Zealand scrutinized 185 prospective studies and fifty-eight clinical trials that had examined the link between fiber intake and health in 4,635 people over a ten- to twenty-year period. The results are astounding. Higher fiber intake was associated with lower body weight, blood cholesterol, blood pressure, risk of heart attack, stroke, bowel cancer, and type 2 diabetes. The data indicate that shifting 1,000 people from a low-fiber diet of less than 15 grams, which is common in the Western world, to a high-fiber diet of more than 25 grams would prevent six cases of heart disease and thirteen premature deaths from all causes. There was a clear dose-response relationship, indicating that as far as fiber goes, more is better.

Exactly why fiber should have such benefits is not totally clear, but the effects are likely multifactorial. Fiber prevents bile acids from being reabsorbed and being converted into cholesterol, it speeds potentially toxic compounds through the colon, reduces hunger by providing bulk, and serves as food for beneficial bacteria in the gut that crank out short-chain fatty acids linked to colon health.

Now let's cut to the chase. How do you get at least 25 grams of fiber? An apple has 4 grams, a banana, a carrot, and a handful of almonds each have 3, half an avocado has 6, two slices of whole grain bread have 4, a cup of cooked lentils has 15, half a cup of rolled oats has 9, and the winner is . . . Fiber One cereal with 14 grams in half a cup. But be careful with the current versions of graham crackers. Not much fiber and lots of sugar. Also, think twice about the ultra-low-carb diets.

KETO SURPRISES

No Bread. No pasta. No sugar. No cereals. No potatoes. No carrots. No rice. No beer. Few fruits. On the other hand, butter, high-fat cheese, and fatty meats are fine. What we are talking about is the increasingly popular keto diet. What do we really know about a diet that eschews carbohydrates to this extent?

Most keto regimen experimenters are trying to gain the upper hand in the proverbial battle of the bulge by cutting carbs. That idea is not novel. Way back in 1864, William Banting in a "Letter on Corpulence, Addressed to the Public" suggested avoiding "any starchy or saccharine matter which tends to the disease of corpulence." Banting was no medical expert. He was a British undertaker with a girth that would probably have prevented him from fitting into one of his coffins.

After twenty frustrating years of trying everything from Turkish baths to starvation diets, he sought help from Dr. William Harvey, a physician who had been following famed French physiologist Claude Bernard's work on the metabolism of sugar. Harvey advised Banting to give up sugar, as well as bread, butter, milk, beer, and potatoes with the result that the undertaker lost 46 pounds and 12 inches off his waist in just thirty-eight weeks. It was basically Banting's diet that was resuscitated and popularized by Robert Atkins in the 1970s, and it is an extreme version of the Atkins diet that is now promoted by dedicated keto worshippers.

There is little doubt that cutting way back on carbs results in weight loss. The question is why? The body's main source of energy is glucose, generally supplied by starches and sugars in the diet. If consumption of these carbohydrates is drastically reduced, below about 50 grams a day, energy has to be derived from an alternate source. While there are essential dietary proteins and fats, there are no essential carbohydrates. That's because the body can make glucose from proteins and fats. At first, the 65 or so grams of glucose the body needs per day are produced from amino acids, sourced from proteins. But this process itself has a high energy requirement, and furthermore, the body is not keen on using up proteins that are needed to maintain muscle integrity. Fortunately, there is a backup system that can swing into action.

The liver begins to convert fats into ketone bodies, namely beta-hydroxybutyrate, acetoacetate, and acetone. These are then shuttled into the mitochondria, the cells' little energy factories, where they are used as fuel. At this point the body is said to be in "ketosis," with excess ketones being excreted in the urine. Some acetone finds its way into the lungs and is responsible for the characteristic sweet scent of "keto breath."

The usual argument for the more efficient weight loss associated with extremely low-carb diets as opposed to low-fat diets is that they produce a metabolic advantage because a lot of calories are needed to convert proteins to glucose. Not everyone agrees with this, though. Some suggest that ketone bodies either have a direct appetite suppressant effect, or that they alter levels of ghrelin and leptin, the respective appetite stimulating and inhibiting hormones. Then there are those who argue that ketogenic diets succeed simply because they lead to a lower calorie intake. While it may sound seductive that keto worshippers can eat as much steak and butter as they want, they actually end up consuming limited amounts, likely due to the greater satiety effect of protein.

Now for the question of whether the high fat content of a keto diet impacts cardiovascular risk factors. As one would expect, LDL, the "bad cholesterol," does go up, although the increase is mostly in the "large particle" subfraction that is deemed to be less risky. Triglycerides, a significant risk factor, actually decrease on a very-low-carbohydrate diet, as does the body's own production of cholesterol. Levels of HDL, the "good cholesterol," increase. But again, the problem is that there are no studies of people who have followed a keto diet long enough to note whatever effect such a diet may have on heart disease.

Not everyone is "doing keto" to lose weight. Some are seeking better brain function, improved control of blood sugar, help with neurological ailments, or even starvation of tumor cells by limiting their supply of glucose.

The most significant evidence for keto benefits is in childhood epilepsy. Hippocrates, the most famous of the ancient Greek physicians, noted that fasting reduced the frequency of epileptic seizures. That idea was not put to a test until the early twentieth century when clinical trials demonstrated seizure improvement after two to three days of fasting. When it was found that starvation resulted in elevated blood levels of acetone, beta-hydroxybutyric acid, and acetoacetate, these compounds became candidates for the therapeutic effect. These "ketone bodies" form when due to a lack of glucose, as in a low-carb diet, the body has to resort to the use of its fat stores to supply energy. It is the breakdown of fat that yields the ketones that can serve as an alternative energy source for the brain, and their use instead of glucose somehow reduces seizures.

Since epilepsy is a brain disorder that responds to an influx of ketones into the brain, it raises the question of whether general brain function may be improved by a ketogenic diet. When scientists try to answer such questions, they usually first turn to rodents. Mice and rats can be trained to navigate mazes and avoid electric shocks, and indeed they do show an improvement in memory when they are on a ketogenic diet. That doesn't mean much, because, with some notable exceptions, human brains are more sophisticated than those of rodents.

As far as humans go, there is much Internet chatter about improved brain function, with all sorts of theories being proposed. Ketosis reduces symptoms of stress and anxiety by increasing the production of gamma-aminobutyric acid, an inhibitory neurotransmitter. Ketones make nerve cells function better because they are a more efficient fuel for neurons than glucose. Ketones reduce oxidative stress in the brain by increasing production of the antioxidant glutathione. Interesting theories, with no real evidence. There are personal accounts of greater "mental clarity" and alleviation of "brain fog" on a keto diet, but there are no compelling studies to back up the claims.

That hasn't stopped supplement manufacturers from cranking

out a variety of "keto" products, mostly containing beta-hydroxybutyrate, with claims of improved mental function, neuroprotection, increased fat-burning, and optimizing athletic performance, all without any significant supporting evidence. In fact, studies have shown that in professional cyclists, performance is actually impaired with such supplements, and that ketogenic diets turn out to be disastrous for racewalkers.

The scenario is a lot more optimistic for type 2 diabetes, where keto diets can result in greatly improved glucose control and better insulin sensitivity, in some cases even eliminating the need for medication. At least one study has shown better glucose control in type 1 diabetes, along with fewer adverse events.

When it comes to claims about treating cancer, Alzheimer's disease, amyotrophic lateral sclerosis (ALS), and Parkinson's disease, not only is the jury out, there isn't even enough evidence to assemble one. With cancer, the theory is that insulin stimulates the division of cells, and that cancer cells rely on glucose for energy more than other cells, so reducing glucose in the bloodstream results in less insulin production as well as less fuel for cancer cells. An interesting possibility with no clinical evidence.

Keto diets are not without concern. Starting such a diet can lead to headaches, fatigue, irritability, and nausea, symptoms referred to as the "keto flu." Constipation can be a problem due to lack of fiber but surprisingly, cardiovascular risk factors are not increased. Some concern has been raised about high-fat diets altering the microbial composition of the gut in such a way as to favor the formation of inflammatory compounds.

Many keto diets feature an abundance of meat. Not a great idea, given that according to a recent study of some 15,000 adults over twenty-five years, mortality rate increased on a low-carb, high-meat diet. This, however, was not found when carbs were replaced by plant sources of protein. So, for those interested in giving the keto diet a shot, whether it be for weight loss or control of diabetes, the best bet is to follow a version in which protein comes from vegetable sources, supplemented with fish or poultry. And drenching everything with butter is not a requirement.

Conclusion: There are numerous studies published over the last twenty years that have compared low-fat diets to low-carb diets with the overall conclusion that the low-carb diets are more effective in terms of weight loss, at least in the short-term. But studies have also shown that in the long-term, weight loss is independent of the composition of fat, carbohydrate, and protein in a diet as long as caloric intake is reduced. Surprisingly, a keto diet does not seem to increase cardiovascular risk factors although studies longer than six months are lacking. Impressive results have been noted in controlling type 2 diabetes with such a diet.

It is, however, interesting to note that the longest life expectancy in the world is on the Okinawa Islands of Japan, where calories from carbohydrates in the diet outnumber those from protein by ten to one. Of course, their carbs don't come from sugar or refined grains, but rather the likes of sweet potatoes, bitter melon, and soybeans. It seems that nobody has a monopoly on the "healthiest" diet.

THE MUCH-MALIGNED EGG

Alfred Hitchcock, who struck fear into the hearts of movie audiences with *Psycho* and *The Birds* had a personal terror of his own. The famous director suffered from ovophobia, the fear of eggs. In a 1963 interview, he claimed to have never tasted one because he was repulsed by the sight of an egg yolk breaking and spilling its yellow liquid. Hitchcock was not alone in looking at eggs with trepidation in the 1960s. But it wasn't the look of the egg that was terrifying people. It was what the yolk contained. Cholesterol!

This naturally occurring chemical was first linked with heart disease in the early twentieth century by Russian medical researcher Nikolay Anichkov, who found it to be a major component of deposits in the coronary arteries. Discussions about cholesterol entered the public forum in the 1960s when data emerged from the famous Framingham study that had been monitoring the health status of a number of volunteers in the small Massachusetts town since 1948. Higher levels of cholesterol in the blood were associated with an increased risk of heart disease! Other scientists discovered that feeding cholesterol to test animals increased their blood levels and that in people, the amount of cholesterol consumed in the diet correlated with the risk of atherosclerosis, the hardening of arteries.

All of that seemed to be enough evidence for the American Heart Association to portray cholesterol as a villain and recommend that people should consume no more than 300 milligrams per day. Eggs were to be limited to three a week, but there were also dissenting voices. The animal-feeding studies had been mostly carried out with rabbits, herbivores that do not normally have cholesterol in their diet and likely respond differently from humans. The validity of linking dietary cholesterol to heart disease was also questioned since diets high in cholesterol tend to be high in saturated fat, which boosts cholesterol synthesis in the body. So, while it was clear that blood levels of cholesterol were associated with heart disease risk, it was not at all clear that cholesterol in food, and specifically eggs, was a major culprit.

The situation began to be clarified in 1999 with results from the Health Professionals Follow-Up Study and the Nurses'

Health Study. Harvard researchers had followed over 100,000 participants for at least twelve years, monitoring the incidence of cardiovascular disease. Based on food frequency questionnaires, they concluded that that "consumption of up to one egg per day is unlikely to have substantial overall impact on the risk of coronary heart disease or stroke among healthy men and women." They did, however, note an apparent risk associated with higher egg consumption among diabetic participants and noted that further research was warranted. Finnish scientists took up the challenge in 2015 and found that in middle-aged and older men, if anything, increased egg intake was associated with a decreased risk of type 2 diabetes.

A collaborate effort in 2017 between researchers in China and at Harvard examined all the publications in the scientific literature that had investigated the relationship between egg consumption, stroke, and cardiovascular disease. Up to one egg a day was not associated with any increase in risk, but there were still questions about whether this applied to diabetics.

Also in 2017, University of Connecticut scientists examined the relationship between egg intake and HDL cholesterol and LDL cholesterol. Since cholesterol is insoluble in water, it is transported around the bloodstream linked to lipoproteins. These come in two basic forms, high-density lipoproteins (HDL) and low-density lipoproteins (LDL). The former is referred to as "good," because it actually removes extra cholesterol while the latter is "bad," because it causes cholesterol to be deposited in arteries. Blood tests from thirty-eight participants who had consumed from one to four eggs per day for four weeks showed improved HDL function with increasing egg intake. While LDL cholesterol also increased, it was the large particle LDL type, which is not the problematic variety.

Taking these studies into account, the American Heart

Association has dropped its reference to egg restriction and the latest edition of *Dietary Guidelines for Americans*, compiled by a body of experts, no longer makes any mention of the 300 milligram daily limit of cholesterol. *Canada's Food Guide* simply includes eggs as a good protein source. A massive study in China of over half a million people in 2018 further exonerated eggs with the conclusion that "among Chinese adults, a moderate level of egg consumption, up to one egg per day, was significantly associated with lower risk of cardiovascular disease, largely independent of other risk factors."

Case closed? Nope. Yet another study on the relationship between egg consumption and cardiovascular disease appeared in 2019 in the *Journal of the American Medical Association* (JAMA) and came to the conclusion that "higher consumption of dietary cholesterol or eggs was significantly associated with higher risk of cardiovascular disease." This triggered headlines ranging from the alarmist "Higher Egg and Cholesterol Consumption Hikes Heart Disease" to the absurd "Eating Three Eggs a Week Can Kill You." Unsurprisingly, that one was from the animal rights organization People for the Ethical Treatment of Animals (PETA).

The JAMA study caused quite a commotion both among the public and in the scientific community because it flies in the face of the numerous studies over the last few decades that have concluded that dietary cholesterol has little impact on blood cholesterol as well as several that have looked at egg consumption specifically and found eggs to be innocent of any nutritional crime. Does this new study provide enough oomph to once again make eggs a pariah? I think not.

Researchers pooled data from six separate studies that had recorded incidents of cardiovascular disease and death rates in some 29,000 people over an average of seventeen years. All the subjects filled out food frequency questionnaires at the beginning of each trial. After calculating the average consumption of cholesterol, the researchers alleged that more than 300 milligrams a day is associated with an increase in cardiovascular disease. How big an increase? Out of a hundred people consuming 300 milligrams a day, three would be expected to eventually experience a cardiovascular event. Whether the cholesterol comes from eggs, meat, or dairy doesn't matter.

The headlines that screamed of three or four eggs a week being risky were based on the assumption that it would take this number of eggs to put the daily average cholesterol intake over the 300 milligrams limit after having estimated the amount contributed by meat and dairy. In other words, if no other animal products were consumed, given that an egg yolk contains about 180 milligrams cholesterol, a dozen eggs could be eaten a week without going over the 300 milligrams daily limit.

How is it, though, that this time researchers found a link between dietary cholesterol and heart disease when many other studies did not? They claim a better job of teasing out the relevant data by controlling for confounding factors such as smoking, physical activity, alcohol intake, and other dietary factors. Maybe. Even if this were so, a dark cloud still hangs over this study. All the conclusions are based on food intake as determined by questionnaires filled out only once, with the assumption being that the subjects then followed the same diet for decades. This is highly suspect. How many of us have not changed our diets over that period of time?

Suppose that back in the 1980s, when these six studies got underway, some people decided to limit egg intake because of the publicity at the time about blood cholesterol being a causative factor in heart disease and eggs being a rich source of cholesterol. Then evolving research began to show that blood cholesterol was actually a function of the specific types of fat consumed and that dietary cholesterol was a minor player. As a result, our subjects began to worry more about fat in meat, lost their fear of cholesterol in eggs and downed them on a regular basis. They never developed any sort of cardiovascular disease. In the study in question, these subjects would fall into the category of no cardiovascular disease thanks to low dietary cholesterol, despite actually not being a low consumer.

Conversely, consider people who ate lots of eggs when the questionnaire was initially administered but were later diagnosed with high blood cholesterol and decided to cut out eggs and red meat, replacing these with low-fat foods that featured lots of unhealthy carbohydrates. If eventually they suffered a cardiovascular event, they would be deemed a victim of high dietary cholesterol even though for most of the duration of the study, little cholesterol was consumed.

Where does all of this leave us? Individual studies can always be criticized, especially when it comes to using food frequency questionnaires, which have reliability issues. There is just no compelling evidence for suggesting that moderate egg consumption, in the range of five eggs a week, is harmful. We can therefore leave ovophobia to discussions of Alfred Hitchcock's psyche.

PUTRID INTESTINAL GUCK?

"People are being sickened by the unelimintated filth that forms a sticky coating on the walls and in the folds of the intestines and releases poisons of putrefaction into the bloodstream." You would think that bit of info comes from one of the numerous books or blog posts that currently promote a plethora of schemes to "detox" the body with various colonic cleanses. But no, that warning dates back to the first decades of the last century, an era haunted by the prospect of intestinal "autointoxication" caused by constipation.

No one played a larger role in promoting the idea that all disease begins in the bowel than Dr. John Harvey Kellogg of cereal fame. A trained physician, Kellogg latched onto the work of Russian zoologist Élie Metchnikoff, who had proposed that toxic bacteria in the gut play a role in aging and that their effects could be countered by "good" lactic acid–producing bacteria. The reason that Bulgarian peasants had impressive longevity (never actually documented) was due to their consumption of yogurt! Kellogg concluded that cleansing the colon of toxic bacteria and replacement with "protective" germs was the key to health. He even designed an enema machine to purge the bowel prior to introducing yogurt both orally and rectally. While it is dubious that his treatments had a therapeutic effect, Kellogg was a pioneer in recognizing the importance of the microbiome, a huge area of research today. Kellogg's cereals were also designed with health in mind, their high fiber content acting like "little brooms" to sweep out the nasty immobilized feces that gave rise to toxins.

The specter of autointoxication was a boon to the sellers of laxatives. "What would be more repulsive than to have one's intestines filled with rotting, foul-smelling, undigested food matter?" asked one advertiser. Products containing castor oil, Epsom salts, and calomel flooded the market along with enema devices, acidophilus supplements, rectal dilators, and internally applied electrical colon stimulators. Into this quagmire stepped phenolphthalein, destined to become the most famous laxative of them all. Yes, the same phenolphthalein that you likely used as an acid-base indicator in your high school chemistry class, colorless in an acid solution and bright pink in base.

Phenolphthalein is a synthetic compound first made in 1871

by Nobel prize–winning German chemist Adolf von Baeyer at a time when, stimulated by William Henry Perkin's accidental discovery of mauve, chemists were hot on the trail of synthetic dyes. It didn't work out as a dye but turned out to be very useful as an acid-base indicator. And then in 1900 came another serendipitous discovery.

Hungarian white wines were in short supply due to a poor grape harvest, and some cheaper imported wines were apparently being sold as authentically Hungarian. The government had the idea of somehow marking real Hungarian wine with a chemical that would not alter its color or taste but would allow for a test to be carried out for its presence.

Phenolphthalein was proposed as a suitable additive. An authentic white wine would then turn pink when a base was added, while an impostor would not. Obviously, the safety of adding phenolphthalein would have to be explored, and chemist Zoltan Vamossy of the Pharmacology Institute of the University of Budapest was asked to look into the chemical's toxicological profile before any plan to add it to wine was put into practice. After animal tests indicated the compound was harmless, Vamossy and a colleague did something that was quite acceptable at the time but would be out of the question today. They became their own guinea pigs and took small doses of phenolphthalein to test for its effects on humans.

Those effects became quickly evident. Both men experienced diarrhea and as Vamossy later recounted, "I had discovered a laxative of great merit." He went on to organize clinical trials with pediatricians reporting they "preferred phenolphthalein to the troublesome castor oil because it was mild in action and pleasant to take." That's was all the German pharmaceutical industry needed to start spewing out products with names like Purgatin, Purgolade, and Laxine.

Enter Max Kiss, a Hungarian-born New York pharmacist who had followed the trials and tribulations of phenolphthalein in his homeland and had the idea of making the product more palatable. After trying different formulations, he hit upon blending the laxative with chocolate. In 1906, the "excellent laxative," Ex-Lax, was born! "A treat instead of torture" boasted an ad. The product became wildly popular, although there were concerns that Ex-Lax would be mistaken for candy by children with disastrous results. There were also stories about mischievous students offering "chocolates" to their teachers hoping for an early end to class.

Although by the 1940s the notion that the majority of people were "dragging through life with foul breath, dulled minds, and sluggish muscles due to chronic constipation" faded, phenolphthalein laxatives remained popular until they were yanked in 1997 due to some questionable animal studies purporting the compound was carcinogenic. Ex-Lax still exists today, but phenolphthalein has been replaced by sennosides derived from the senna plant.

Unfortunately, the idea that much of human misery is caused by putrid intestinal guck that needs to be purged is still touted by some "alternative" practitioners. Too bad there is no solution for mental constipation.

MICROBES IN THE GUT

"Eat your fruits and vegetables!" A common refrain from frustrated parents as they watch their kids disdainfully play with the peas and carrots on their plate. The emphasis on fruits and veggies is sound, based on a legion of studies demonstrating that populations with mostly plant-based diets are less prone to cardiovascular disease, diabetes, obesity, and cancer. Why this should be so is a matter of debate. Is it a matter of what is eaten or what is not eaten? Are the benefits due to some naturally occurring substances in plants, with antioxidants, vitamins, and fiber usually being placed on the pedestal? Or the negative effects of refined grains, sugar, or meat? Or is it a combination of the two?

Dietary supplements containing the supposed beneficial substances in plants have been extensively hyped by marketers but have generally failed to deliver the goods. However, in the last few years, another candidate purporting to plug the holes in the mysteries of the relationship between diet and health has appeared, stimulating vigorous research and prompting both reliable and overzealous media accounts. I'm talking about our "microbiome."

Our skin, mouth, and digestive tract are hosts to somewhere between thirty and fifty trillion bacteria of hundreds of species, collectively referred to as the "microbiome." Actually, there are somewhat more bacterial cells in our body than human cells, but luckily, our appearance is dictated by the human cells. Prevailing opinion had been that we all have roughly the same composition of bacterial species and that aside from helping to break down dietary fiber during digestion, the bacteria do not interact with the body. Recent research indicates that this is far from being the case and that the specific variety of microbes we harbor may well be instrumental in shaping our risk for disease.

For example, microbiologists have found a significantly greater intestinal microbe diversity in young villagers in the African country of Burkina Faso than in Italian children. The Africans' diet is based mostly on whole grains like millet and sorghum while the Italians eat a Western diet with refined grains and simple sugars. It turns out that the more diverse African microbiota produce twice the amount of short-chain fatty acids, with butyric being a prime example. These products of fiber digestion in the colon are small enough to diffuse through the mucus layer lining the intestine into the circulation and can calm low-grade inflammation, a condition that many researchers believe drives a number of the chronic "Western" diseases.

On the other hand, microbes can also stimulate inflammation if their hunger for fiber isn't satisfied. They will then start feeding on the mucus lining of the gut, eroding it and allowing bacterial breakdown products called endotoxins to enter the bloodstream. These are perceived as a threat by the immune system, triggering an inflammatory response aimed at eliminating them.

What is the takeaway message here? That increased fiber intake leads to greater diversity of bacteria in our gut, which enhances short-chain fatty acid production, and at the same time helps maintain the mucus lining of the gut. The result is a reduction in what may be described as dangerous simmering inflammation.

The exact nature of the bacteria in our gut may even be important in determining the risk of other conditions, such as Parkinson's disease. Parkinson's is characterized by a lack of dopamine stimulation of nerve cells because the action of dopamine, a neurotransmitter, is impeded by the buildup of clumps of a protein known as alpha-synuclein, possibly initiated by a signal from gut bacteria. Mice bred to have germ-free intestines develop Parkinson's-like symptoms upon transfer of gut microbes from Parkinson's patients but not with microbes introduced from healthy people. Further research may reveal which bacteria specifically are linked with Parkinson's and whether they can be crowded out by the introduction of other bacteria, so-called "probiotics." But yogurt is not going to do it. Neither will it prevent depression, despite the eye-catching headlines like "Feeling Depressed? Eat Yogurt Rich in *Lactobacillus.*"

The study that generated this seductive headline certainly did not demonstrate that eating yogurt can alleviate depression. What it showed was that stressed mice will struggle more vigorously in a tank of water if they are supplemented with *Lactobacillus reuteri*, bacteria that are used to make yogurt.

When a mouse is placed into a tank of water without any chance of escape, it will at first vigorously swim around trying to find a way out, but within a few minutes, it will realize its hopeless state and sort of float, moving its legs just enough to maintain balance and keep its head above the water. The time it takes for "behavioral despair" to set in is said to be a measure of the animal's mood. This is essentially based on the observation that treatment with antidepressants increases the time the animal struggles; in other words, it stays hopeful for a longer period.

In the study in question, researchers stressed mice by exposing them to noise, strobe lights, crowded housing, and restraint in conical tubes. As if that weren't enough, the cages were frequently tilted, and the mice's bedding wetted. These were stressed animals! And when they were subjected to the forced swimming test, they quickly gave up the struggle. This is interpreted as the mice being depressed. The gut bacteria of the "depressed" mice were then analyzed and found to have fewer lactobacilli than unstressed mice. Treating the stressed mice with these bacteria increased the time they actively struggled in the water, supposedly because they were more hopeful of escape and less "depressed."

Never mind that humans are not giant mice. "Behavioral despair" in trying to escape from a water tank hardly equates to depression in people. Furthermore, the mice were treated with 2 billion colony-forming units (CFU) per day, which taking body weight into account, would require gallons of yogurt for people. You can see why I get depressed when I see headlines about treating depression with yogurt.

A CULTURE LESSON

Genghis Khan's armies supposedly lived on it, Élie Metchnikoff thought it was responsible for the long life of Bulgarian peasants, and Dr. John Harvey Kellogg pumped it into his patients' bowels. Today, shelves in the dairy aisle are brimming with dozens of varieties promising not only great taste but good health. It's yogurt! Labels on packaging clamor about their contents being low-fat or sugar-free or antioxidant-rich or crammed with live cultures or all of the above. And then there is the ever-popular Greek yogurt.

Basically, yogurt is milk soured with lactic acid produced by the action of certain bacterial enzymes on lactose, the sugar naturally present in milk. The acid changes the structure of casein, the major protein present in milk, causing it to form an insoluble curd that is then suspended in a liquidy portion called whey. The bacterial action also breaks down any remaining lactose into its components, namely glucose and galactose, meaning that people with lactose intolerance will have a lot less trouble with yogurt than with milk. Lactose intolerance is characterized by the lack of lactase, an enzyme in the digestive tract that breaks down lactose into its absorbable components. If the enzyme is absent, lactose travels to the colon, where it can trigger diarrhea. The colon also harbors bacteria capable of digesting lactose, producing gas in the process. The result is abdominal discomfort.

In all likelihood, yogurt was an accidental discovery made when naturally occurring bacteria invaded milk, possibly from goatskin used to make bags for carrying milk. Pliny the Elder in the first century AD wrote that "barbarous nations" knew how "to thicken milk into a substance with an agreeable acidity." These barbarians had undoubtedly learned that fresh yogurt could be made by adding a bit of yogurt from a previous batch. Indeed, this is just the way yogurt is made today. A bacterial culture is added to milk that has been heated to kill off any undesirable bacteria. Greek yogurt is thicker than regular yogurt because much of the whey is removed. Traditionally this has been accomplished by straining through cheesecloth, although today some processors use a centrifuge. Another route to Greek-style yogurt involves adding milk protein concentrate and thickeners such as pectin or inulin to regular yogurt.

Removal of the whey has some nutritional benefits. Since the product is more concentrated, it has more protein, about 15 grams per serving compared with 9 grams for the conventional variety. This can help with weight control because protein is filling. Since the straining process removes a lot of the soluble carbohydrates, Greek yogurt has only about half the carbohydrates found in regular yogurt. Some calcium is lost in the whey, but Greek yogurt is still a good source of the mineral with a serving supplying about 20 percent of the recommended daily intake. Keep in mind that if it's made from whole milk, any yogurt will be high in saturated fats, so it is best to stick to the versions made from low-fat or skim milk. And to ones that are made without added sugar.

Now for the downside of Greek yogurt. While it may be more friendly to our health, it is not so friendly to the environment. There are two basic issues. One, Greek yogurt generates large amounts of acidic whey that have to be disposed of, making for an environmental challenge. The whey can't be dumped into water systems because when it biodegrades, it uses up a lot of the water's dissolved oxygen, leading to the destruction of aquatic life. Some of the whey can be blended into feed and fertilizer or potentially be used to provide protein for infant formula and body-building supplements, but more whey is produced than can be economically used. The second problem is that it takes about four times as much milk to make Greek yogurt compared to regular yogurt. That's a problem because milk production itself takes a toll on the environment, due to the production of large amounts of greenhouse gases. Whenever the required fertilizers, pesticides, and feed are manufactured and transported, carbon dioxide is released. Producing cows is an energy-intensive process in the first place, consuming about 50 percent more energy than pork or poultry production. And a lot of energy is used when dairy animals and their calves are eventually slaughtered for meat.

Dairy farms also use a great deal of electricity because of the machinery used to milk the cows, cool the milk, and heat the water needed to wash needed equipment. Processing, transport, and packaging the milk also requires energy. Then there is the cows' own emissions of methane, a more potent greenhouse gas than carbon dioxide. On top of it all, there is the issue of nitrous oxide, another greenhouse gas, being released from ammonium nitrate fertilizer.

Some people argue that organic milk is the way to go because eliminating fertilizers and pesticides saves energy. Also on organic farms, cows graze on clover-based pastures, which is beneficial since clover can take up nitrogen from the air and convert it into compounds in the soil that can be used as fertilizer. Furthermore, organic herds forage more than conventional cows, suggesting that less feed such as soy, which requires an energy input for growing, needs to be imported for the production of organic milk.

What about the nutritional difference between organic and conventional milk? Many newspapers featured headlines along the lines of "Fresh Research Finds Organic Milk Packs in Omega-3s," sending consumers scurrying to the organic aisle in the supermarket. Omega-3 fats have developed an aura of health in spite of murky evidence, but it is true that milk from cows feasting on grass and clover contains more omega-3 fats than milk from grain-fed animals.

However, the omega-3 fat here is alpha-linolenic acid (ALA), which is not the one that has been linked with health benefits from eating fish. That, though, is hardly the point. Press reports hailed the finding of "60 percent more omega-3 fats in organic milk." Whoaaa! Talk about a misuse of numbers. The organic milk had 32 milligrams of omega-3 fats per 100 grams of milk versus 20 milligrams for conventional. Indeed, a 60 percent difference, but totally inconsequential! When we talk about the benefits of omega-3 fats, if there actually are any, we are talking about needing hundreds, if not thousands of milligrams per day. As mentioned, there are benefits to organic milk, but touting it as a source of omega-3 fats is based on a nonsensical milking of the data.

THE PROS OF PROBIOTICS QUESTIONED

I drink kefir almost every day. I like the slightly tangy flavor and the hint of carbonation. I'll admit, though, that the possibility of health benefits has not escaped my attention. Kefir is made by culturing water or milk with "kefir grains," which are not grains at all but small bits of a mixture of bacteria and yeasts. The beverage is considered a probiotic, defined by the World Health Organization as "live microorganisms, which when administered in adequate amounts confer a health benefit on the host." The health benefits that have been attributed to probiotics include improvement of irritable bowel syndrome, reducing the risk of diarrhea as a consequence of antibiotic therapy, better immune function, and even improved mental health. Such discussions center on the potential effects of probiotics on our microbiome, that incredibly complex mixture of microorganisms that populate our gut.

Irritable bowel syndrome (IBS) is a catch-all term for a collage of symptoms that include diarrhea, constipation, bloating, and abdominal pain. Although no clear cause has been identified, some studies have shown that IBS sufferers have a different composition of microbes in their feces than people who are not afflicted. This has prompted research into the treatment of the condition with probiotics, with some evidence of success. Delivering potentially beneficial bacteria to the gut is challenging given that the acidity of the stomach does not make for a friendly environment for bacteria. The usual solution is to pack the microbes into a capsule that traverses the stomach and then dissolves in the alkaline medium of the gut. A couple of studies using *Bifidobacterium infantis* 35624 (found in "Align") have indeed shown a greater reduction in bloating and abdominal pain than with placebo.

Perhaps the greatest focus with probiotics has been on their potential to counter the side effects of antibiotics, mainly diarrhea. Antibiotics are of course extremely useful, but they are rather indiscriminate in their antimicrobial effect, wiping out some "good" bacteria as well as those that cause disease. The "good" bacteria play a role in keeping some of the bacteria that can cause diarrhea in check by competing for the available food supply. If their number falls, the unchecked "bad" bacteria, *Clostridium difficile* being a classic example, multiply and cause misery. In theory, reestablishment of a healthy microbiome should alleviate the problem. Although many probiotic supplements claim to do exactly that, evidence is rather underwhelming. A couple of recent studies from the Weizmann Institute of Science in Israel show that adjustment of the composition of the microbiome with a view towards benefits is a very complex matter.

First, most studies assume that the composition of the microbiome is reflected by the composition of bacteria in fecal matter. This may not be the case. Immunologist Eran Elinav and his team enlisted twenty-five healthy volunteers who were willing to undergo endoscopies and colonoscopies in order to map their baseline microbiomes at different parts of the gastrointestinal tract. Kudos to them! As it turns out, the microbial content of stool samples only correlated partially with that in the gut, suggesting that fecal matter is not a good proxy for gut bacteria.

Next, for four weeks, some of the volunteers took a commercially available probiotic containing eleven strains of bacteria while others took a placebo. Then they were all scoped again. Dedicated volunteers indeed! Surprisingly, to a greater or lesser extent, the probiotics were mostly pooped out. They tended to be retained more by volunteers who had lower levels of the probiotic strains in their natural microbiota in the first place, suggesting that successful probiotics may have to be custom designed by taking into account individual microbiomes.

Even more perturbing were the results seen after the volunteers were treated with antibiotics for a week. As expected, this altered their microbial community significantly. Some subjects were then given a daily dose of probiotics, some a placebo, and some had the pleasure of experiencing transplants of their own feces that had been collected before the antibiotic treatment. Once more, surprisingly, the probiotic subjects did not show the expected benefits. It took five months before their microbial community was reestablished! The control group saw theirs reestablished in just three weeks, while the fecal transplant subjects fared the best, their microbiome recovering in just a day! But as they say, don't attempt this at home.

Speaking of fecal transplants, this procedure is being increasingly used to treat *Clostridium difficile* infection, which is acquired in hospitals with alarming frequency. In this case, the transplant makes use of fecal bacteria from a healthy donor. Dr. Lawrence Brandt, a gastroenterologist at Montefiore Medical Center in New York, came up with this novel, though admittedly unconventional, approach. In one highly publicized case, he mixed stool samples from a patient's husband in saline water and deposited little chunks of this matter every 10 centimeters along the woman's colon. This "fecal colonoscopy" resulted in the almost immediate resolution of symptoms, in all likelihood due to the restoration of a healthy balance of microflora in the gut. This doesn't sound particularly appealing, but protracted diarrhea is no pleasure either.

Interestingly, while transplanted fecal matter is screened for human immunodeficiency virus (HIV) and hepatitis C, some researchers believe the donor should also be screened for mental illness because of accumulating evidence that the composition of the human microbiome may be linked to mental status.

Bacteria in the gut are known to produce neurotransmitters such as dopamine, norepinephrine, and serotonin that can make their way to the brain either via the vagus nerve or the bloodstream. It seems that the microbiome of people with depression differs from that of people who do not have mental problems. Interestingly, when fecal samples from people with depression are transplanted into rats, the rodents begin to show signs of depression and anxiety. In a study of twenty-two men treated for a month with a placebo or with a specific bacterium, *Bifidobacterium longum*, levels of the stress hormone cortisol were found to be lower with the bacterium, again demonstrating that bacteria in the gut can affect activity elsewhere in the body. Perhaps in the future, "psychobiotics," specific bacterial compositions, may be used to treat stress or even mental illness.

The problem is that for now, nobody knows what the composition of a "healthy" microbiome is. Indeed, it may not be the same for everyone. All that can be said is that "some probiotics seem to be useful for some conditions some of the time." There is better news on the prebiotic front. Prebiotics are nondigestible food components that promote the growth of beneficial microbes in the intestines. A study with forty-five volunteers showed that taking a commercial supplement of galactooligosaccharides (under the brand name Bimuno) for three weeks resulted in a reduction of the stress hormone cortisol. Maybe I should add that to my daily regimen of kefir given that trying to tease reliable information out of all the probiotics studies is stressful.

EAT LIKE THE HADZA

They forage for berries, tubers and honey. They hunt birds and game with poisoned arrows. Porcupine is a favorite and baboon brain is considered a delicacy. They eat the way our ancestors ate before agriculture was introduced some 10,000 years ago. They have virtually no possessions and no fixed place to live. They are the Hadza people of Tanzania, one of the world's last remaining hunter-gatherer tribes. And they are of great interest to researchers because they appear to experience no diabetes, colon cancer, colitis, or Crohn's disease. Obesity is unknown. One theory is that these conditions, which are rampant in the developed world, may be the consequence of a shifting balance in gut bacteria as a result of our modern diet.

Up to a thousand different varieties of bacteria inhabit our intestines, contributing up to two kilos to our body weight. Like other living organisms, bacteria eat and defecate. Their food is whatever we throw in their direction. They will digest the components we can't, and then reward us for feeding them by releasing a variety of chemicals that can contribute to our health in their poop. Among these are vitamins B12 and K as well as short-chain fatty acids that help keep our gut in good shape.

One way to gain insight into the connection between gut bacteria and health is to look for differences in the microbiome of dramatically different populations, and there can hardly be two populations as different as urban Italians and the Hadza. It turns out that bacteria in feces are a good reflection of bacteria that reside in the gut, and when researchers compared the bacteria in the poop of the Italians and the Hadza, they found a 40 percent greater diversity of microbes in the Hadza output.

Interestingly, the feces of the Hadza show an absence of *Bifidobacterium*, species that are thought to be important in the Western diet and ones that are commonly included in probiotic products. Interestingly, the Hadza had lots of *Treponema*, a type of bacteria that in Western populations is associated with diseases such as Crohn's and irritable bowel syndrome, yet the Hadza do not suffer from these conditions. Another noteworthy finding was that Hadza men and women had significantly different bacterial populations, in all likelihood, due to different diets. The men eat more meat and honey, and the women eat more fibrous tubers.

In order to decipher the nuances of the relationship between our microbiome and disease, it would be of interest to know if the microbiome can be altered by a change in the diet. That is just what Dr. Tim Spector, professor of genetic epidemiology at King's College, London, wanted to find out. For three days, he would live and eat the Hadza way and monitor the bacterial composition of his poop. He feasted on tubers, wild honey, Kongorobi berries, the fruit of the baobab tree, and roasted porcupine. He accompanied the Hadza on their hunting and foraging forays to expose himself to all the microbes that the Hadza normally encounter, and he refrained from washing. Lab tests showed a highly significant 20 percent increase in microbial diversity after three days and the presence of African microbes that had not been present before. But on returning to England, his microbial profile quickly returned to what it had been before the experiment. Still, the trial did demonstrate that the composition of the microbiota can be altered, albeit it takes a continuous change in diet to do so.

Of course, we are hardly going to make porcupine a dietary staple, and neither baobab fruit or Kongorobi berries are going to show up in our supermarkets. However, we can eat more fruits and berries and high-fiber foods that should go some way to increasing the diversity of our gut bacteria. We must keep in mind, though, that the Hadza are also exposed to various bacteria from the local soil, from animals, and from lack of Western hygiene. Furthermore, the Hadza do not visit doctors, so we really can't be certain to what extent they are protected from Western diseases.

The intriguing research into the human microbiota has spawned numerous probiotic products ranging from pills and fermented dairy products to various beverages. While probiotics may be of help in some intestinal disorders, there is no evidence that they offer any benefit to healthy people. Probiotics contain only a few types of bacteria and generally in numbers not significant enough to make an impact on the composition of the microbiome even if they survive passage through the acidic environment of the stomach. The hope is that future research may be able to identify the specific bacteria that are beneficial to our health and find a way to deliver them to colonize the gut. Rear entry seems promising. Or we can just move to Tanzania and join the Hadza.

THE FOOD OF THE GODS

Let's get something straight right off the bat. Chocolate is not an aphrodisiac, and it does not cause people to fall in love. On the other hand, it may lift our spirits and perhaps even offer some protection from the damaging effects of high blood cholesterol.

The aphrodisiac story is an ancient one. It goes back all the way to 1519 and the first visit of the Spanish explorer Hernando Cortes to Mexico. Cortes found much to his liking here, in particular the Aztec princess Doña Marina. Apparently, the affection was returned because the princess introduced Cortes to a drink made from the pods of a tree. The Aztecs called the drink *chocolatl*, or "food of the gods." The concoction was also laced with dried chili peppers, and as Doña Marina said, it would "stimulate amorous adventures."

Cortes must have been impressed by the effects because on his return to Spain, he presented Emperor Charles V with a sample of cocoa, as we call the substance today. Within a few years, Europeans were indulging in chocolate and singing its praises. Everyone, except nuns that is; they were forbidden to partake of chocolate's pleasures because of the potential consequences. But alas, chocolate does not have aphrodisiac properties. The myth can be ascribed to the presence of general stimulants, such as caffeine, theobromine, and the newly discovered anandamide in chocolate.

Chocolate actually contains over 300 compounds with imposing names like furfuryl alcohol, dimethyl sulfide, phenylacetic acid, and phenylethylamine. It is this last amphetamine-like substance which has been alluringly labeled as the "chemical of love." People in love may actually have higher levels of phenylethylamine (usually abbreviated as PEA) in their brain as surmised from the fact that their urine is richer in a metabolite of this compound. In other words, people thrashing around in the throes of love pee differently from others.

This observation has stimulated the following thought process: falling in love is associated with higher PEA levels, chocolate contains PEA, therefore chocolate can make us to fall in love. Not so! Blood levels of phenylethylamine do not rise after eating chocolate. It seems that most of this enchanting compound is metabolized during digestion. Furthermore, chocolate isn't even a very good source of PEA. Sauerkraut is far better! But that doesn't make for nearly as good a story on Valentine's Day.

So why are we so infatuated with chocolate? Could it have something to do with anandamide, a compound the brain normally produces to signal pleasure? Indeed, anandamide receptors can be stimulated by foreign substances such as tetrahydrocannabinol, or THC, the active ingredient in marijuana. It bears a chemical similarity to anandamide and therefore triggers pleasurable sensations. Chocolate contains anandamide itself, so should it not have the same effect? Probably not.

The amount of anandamide in chocolate is actually very little when compared with the amount produced naturally by the body. An adult would have to eat more than 10 kilograms of chocolate to get a buzz! Well, maybe a little less. A couple of other recently isolated compounds from chocolate, N-oleoylethanolamine and

N-linoleoylethanolamine, inhibit the breakdown of anandamide and may result in higher blood levels.

There is yet another candidate for the secret behind the appeal of chocolates. Endorphins are a class of naturally occurring substances synthesized in the human brain in response to a variety of stimuli. In general, they have been linked to effects similar to those caused by opium. "Runner's high," for example, has been ascribed to endorphin production. According to some researchers, chocolate stimulates endorphin release. This hypothesis is based on the observation that when volunteers are treated with naloxone, a drug that blocks the effect of endorphins, they get no more pleasure out of eating Snickers or Oreos than from eating celery sticks.

Chocolate of course is also high in carbohydrates, mostly sugar. Numerous studies have shown that carbohydrates increase the levels of an important brain chemical known as serotonin. This substance has decided antidepressant effects; in fact, several common antidepressant medications work by increasing concentrations of serotonin in the brain.

But do we really have to get into the nitty gritty of complex brain chemistry to explain our love affair with chocolate? Can it not be that this combination of flavors, sugar, and fats which melts exactly at body temperature just tastes great? Sure it can. However, there is a diversity of opinion on just which chocolates taste the best.

American chocolate certainly tastes different from Belgian, Swiss, and British varieties, a point that repeatedly gets made in British newspapers, often claims that "the reason American chocolate tastes so terrible is that some bars like Hershey's contain the chemical that gives vomit its smell and taste." They're talking about butyric acid. Indeed, the process that Milton

Hershey introduced in the 1930s for making milk chocolate resulted in some of the fats in milk being broken down and releasing small quantities of butyric acid. Americans liked the taste of Hershey's milk chocolate and competitors attempted to copy the manufacturing process, sometimes adding butyric acid. Although the amount of butyric acid in American chocolate is very small, it is enough to have given rise to the silly claims in the British tabloids.

Butyric acid actually forms naturally in the digestive tract as a result of microbial breakdown of food components. That is the reason it can appear in vomit. Its presence there provokes a negative connotation, but in the digestive tract, it actually has a positive effect. Here, butyric acid seems to prevent the irregular multiplication of cells and also has anti-inflammatory activity. Furthermore, there is accumulating evidence that it may be useful in the treatment of irritable bowel syndrome, a condition that affects roughly 10 to 15 percent of the adult population. It is characterized by abdominal pain or discomfort that occurs in conjunction with altered bowel habits.

Pure butyric acid is not suitable for treatment because of its smell, and in any case, if it is orally administered, it is absorbed in the upper part of the gastrointestinal tract and doesn't make it to the colon. However, if the butyric acid is microencapsulated, it can be administered orally. In one study, sixty-six adults with irritable bowel syndrome were randomized in a double-blind, placebo-controlled study to receive either 300 milligrams per day of microencapsulated butyric acid or placebo. After four weeks of treatment, there was a significant decrease in gastrointestinal problems in the group that received the butyric acid. A second trial showed similar results with microencapsulated sodium butyrate. No significant side effects were noted in either trial.

Butyrate concentration in the colon can also be increased with a diet rich in prebiotics, indigestible food components that bacteria in the gut can use to produce butyric acid. Resistant starch, found in whole grains and seeds as well as green bananas, is particularly efficient in producing butyrates. The increased intake of highly processed grains in the Western diet, and the resulting decrease in short-chain fatty acids in the colon, may be a reason for an increased incidence of irritable bowel disease. Another reason to eat those fiber-filled whole grains and veggies.

Nobody is suggesting that American chocolate be eaten for its butyric acid content, but research does suggest that chocolate may actually have some redeeming nutritional features. Although chocolate is of course high in fat, the specific types of fat it contains does not seem to raise cholesterol. Then there is the presence of polyphenols. These are the same compounds that have received a great deal of publicity in connection with the supposed benefits of red wine. Laboratory studies have shown that they can prevent the oxidation of LDL cholesterol (the "bad cholesterol") to a form that damages arteries. A typical chocolate bar actually has the same phenolic content as a glass of red wine; the darker the chocolate, the more phenolics it contains.

Although the polyphenol evidence may not be enough to suggest greater indulgence in chocolate, everyone agrees that smelling chocolate is harmless enough. In fact, it may be beneficial. A study at Yale University has shown that students exposed to chocolate smell while studying for an exam can recall the material better if they are also exposed to chocolate smell while writing the exam. Unfortunately, university regulations do not allow chocolate bars on exam desks. That's because there have been cases of using the inside of wrappers for purposes other than sniffing.

POND SCUM FOR HEALTH?

Some unflatteringly call it pond scum, but to its promoters, it is a "miracle nutrient." Scientifically, spirulina is the dried version of a simple form of blue-green algae, more specifically classified as cyanobacteria. Algae are organisms mostly found in aquatic environments that share a common feature, the ability to photosynthesize. That means they contain chlorophyll, a molecule that can absorb sunlight and use that energy to convert water and carbon dioxide into carbohydrates and oxygen. Algae are critical to life, producing roughly 50 percent of the oxygen in the atmosphere. Although they resemble plants, they are distinct entities since they lack true roots, stems, and leaves. They can be single-celled cyanobacteria, more complex picoplankton, measured in millionths of a meter, or giant kelp that can grow up to 60 meters.

The cells of some cyanobacteria are spiral-shaped, hence the term *spirulina* from the Latin *spiru* for tiny spiral. On the surface of some lakes or ponds, these bacteria aggregate into a mass that can be scooped up and dried, forming a greenish edible cake. Historically, spirulina is known to have been consumed by the Aztecs, who even offered it to the Spanish conquistadores.

Today, these cyanobacteria are cultured in ponds for conversion into a powder that is sold as a nutritional supplement, often with exuberant claims. The powder is about 60 to 70 percent protein and contains a variety of vitamins and minerals. On a gram per gram basis, it is very nutritious, but at the doses recommended, usually in the range of 1 to 4 grams, neither the protein content nor the other nutrients make a significant contribution to the diet. Though it's promoted as a source of calcium, spirulina actually contains only about 5 milligrams in a 4 gram serving. By comparison, a cup of broccoli has 45 milligrams and will also deliver other nutrients comparable to spirulina at a fraction of the cost. Sometimes spirulina is promoted as a source of vitamin B12 for vegans, but the form of B12 that cyanobacteria contain is biologically inactive in humans.

Spirulina is likely to be safe, but since it is regulated as a food, not as a drug, there is no verification that a product contains what the label declares. Some varieties of blue-green algae produce highly toxic compounds called microcystins that can lead to vomiting, rapid heartbeat, nausea, and liver damage. These algae can grow alongside spirulina in natural environments and pose a risk of contamination. However, most spirulina destined for supplements is grown in a controlled environment and is hopefully tested for toxins. Various human illnesses have been reported as a consequence of contact with toxic cyanobacteria through recreational activities such as canoeing or swimming through blooms of blue-green algae.

Now for the hype. Some promoters claim that spirulina may prevent, treat, or cure conditions such as hypertension, diabetes, depression, high cholesterol, cancer, and even Lou Gehrig's disease (ALS). The weasel word "may" does come in handy. Spirulina is also said to boost the immune system and improve liver and kidney function. There are also meaningless assertions about energizing and detoxifying the body and purifying the blood.

When challenged to provide evidence for the claims, promoters point at studies carried out in cell cultures or in animals. Spirulina extracts have shown some anti-inflammatory effects in cells and a reduced risk of atherosclerosis in rabbits fed a high cholesterol diet. A few small-scale human studies have shown a beneficial effect in controlling blood glucose levels in diabetics at a dose of 2 grams a day and improved exercise performance in athletes at 6 grams a day. One study showed that supplements slightly increased hemoglobin levels in senior citizens. None of these studies are compelling.

The major concern is when studies that have little relevance to humans are dredged out to support claims that a product can treat serious conditions. These usually focus on phycocyanin, the blue colorant in spirulina that in laboratory studies can be shown to have antioxidant and anticancer effects. There is nothing special about this, as numerous compounds isolated from various life-forms show such effects in the lab but fail in clinical trials in people.

Where spirulina has been shown to have a practical use is as a supplier of blue food coloring. The green in blue-green algae is due to chlorophyll, while phycocyanin is responsible for the blue tinge. The latter can be extracted to yield a vivid blue that can be used as a "natural" alternative to artificial colors such as Brilliant Blue FCF (FD&C Blue # 1). It can be used in ice cream, puddings, yogurts, and confectionary. When Nestlé pledged to remove artificial colors from Smarties in 2006, there was no alternative to Brilliant Blue, and for about two years there were no blue Smarties. Then researchers hit upon spirulina extract and consumers who had been feeling blue were happy again. So far, surprisingly, nobody has claimed that blue Smarties offer a health benefit due to their phycocyanin content. We're waiting.

COCKROACH MILK

Back in 1981, entomologist Dr. Josef Gregor called a press conference to announce a remarkable discovery. He had bred a novel species of cockroach from which he managed to extract a hormone that, when incorporated into a pill, exhibited amazing properties. It cured conditions ranging from acne and allergies to asthma and arthritis! "Roach Hormone Hailed as Miracle Drug" crowed headlines! Some 175 newspapers went on to feature testimonials attesting to the wonders of the hormone pills.

Subsequently, Dr. Gregor was invited to appear on various television programs where he described that cockroaches were impervious to radiation and that in addition to its curative properties for a plethora of ailments, his pills would offer protection against radiation exposure. It all sounded great, but there was one tiny little problem. There was no Dr. Josef Gregor, and there was no cockroach hormone! Gregor (a name inspired by Kafka's *The Metamorphosis*) was actually Joey Skaggs, a teacher at New York's School of Visual Arts, who relished pulling off hoaxes to show how the media could be duped into reporting nonsensical stories because of a failure to fact-check. And that was decades before the current wave of publicity about "alternative facts"!

Recalling the cockroach hormone episode, I figured a prankster must have been at work when the headline "Scientists Think Cockroach Milk Could Be the Next Superfood" recently scooted across the Internet. Obviously, fact-checking was in order. While the headline was typical clickbait, it was actually spawned by legitimate research.

In 2016, a paper in the *Journal of the International Union of Crystallography* reported some intriguing research about the unique Pacific beetle cockroach (*Diploptera punctata*). Why unique? Because it is viviparous, meaning females give birth to live offspring. The term derives from the Latin *vivus* for "alive," and *parere*, meaning "to bring forth" or "to bear."

While common in mammals, viviparity in insects is rare. The Pacific beetle cockroach does, however, reproduce in this fashion and is of interest to scientists because the embryos get their nutrients from tiny crystals that form from a fluid they absorb from the mother roach. These crystals can be isolated and have been creatively dubbed "cockroach milk" by publicity-seeking headline writers.

The researchers' intent was to study the specific composition and folding pattern of the proteins found in the crystals, since such crystals are rare in living species. They discovered that the proteins were attached to sugars and fats and were extremely tightly packed in the crystalline lattice. A single crystal was estimated to contain three times as many calories as an equivalent mass of dairy milk. Its proteins contained all the essential amino acids and incorporated the necessary fats and carbohydrates needed by a growing embryo. This led to claims that the crystals were a "complete food." Yes, they obviously are for the quickly growing cockroach embryos, but any suggestion that they represent a viable alternative to dairy milk for people is a very, very big stretch.

To start with, milking cows is a lot easier than "milking" cockroaches, and given that some 1,000 roaches have to be sacrificed to get 100 grams of crystals, cockroach milk does not appear to be an economical source of nutrients. There are attempts to isolate the genes responsible for the production of this nourishing liquid, with the hope it can be inserted into the genome of yeast cells that would then crank out the "milk," potentially for human use. Safety would of course have to be addressed.

It should be pointed out that in no way did the researchers claim any sort of superfood status for the crystals. That was a media invention. "Superfood" is a marketing term, not a scientific one. It generally refers to foods that claim to offer an advantage in maintaining health, often based on some sort of study in which animals exhibited a benefit when fed amounts that on a weight per weight basis are greater than can ever be consumed by humans.

The list of superfoods seems endless, ranging from common foods like berries, kale, fish, coconut oil, chocolates, bone broth, beetroot, oats, pomegranate juice, and avocado to the esoteric like chia seeds, goji berries, microalgae, mangosteen, and seaweed. There is nothing wrong with eating these, but the concept of single foods making significant contributions to health is flawed. While there are no superfoods, there are good diets and poor diets. Loading up on fresh produce and curbing processed foods is the way to go.

As far as cockroaches go, you can be awed by their ability to survive radiation and to go for weeks without food or water. The male's knack of attracting females from long distances by wafting his pheromones into the air is also impressive. So is the fact that 100,000 roaches can descend from a single pair within a year.

Amazing creatures, indeed, but don't wait for their "milk" to appear in your grocery store.

HOT DOG FOLLIES

Spelling bees can be fun and educational, but not when they are hijacked to push misleading information. That is exactly what Maple Leaf Foods has done with television commercials that promote its line of processed meat products containing only "simple ingredients with names that you can pronounce." The ads feature youngsters competing in a spelling bee who are stymied by the challenge of spelling "sodium diacetate." They struggle but fail. The message to viewers is that if kids can't spell it, it isn't fit to eat. But should the safety of food components be determined by the spelling abilities of elementary school students? Is "botulin" safe to consume given that spelling it is not be a formidable task?

Sodium diacetate is one of twenty-six substances that Maple Leaf has pledged not to use since they do not meet their criteria of being natural ingredients with names that are easy to pronounce. Take some acetic acid, that's the acid in vinegar, partially neutralize it with sodium bicarbonate, and presto, you have sodium diacetate, a colorless solid that adds flavor to sausages and has antimicrobial properties. When added to food, it performs the same task as vinegar, releasing acetic acid, but as a solid, it is easier to use. While sodium diacetate is to be shunned, vinegar is on the list of simple ingredients that Maple Leaf worships in its quest to avoid selling "fake foods."

Also on the favored list is "cultured celery extract," which was not given its name because the stalks have been exposed to literature or classical music. Rather, the extract has been treated with a bacterial culture that converts the nitrate it naturally contains to nitrite. Guess what! Sodium nitrite is on the list of ingredients the company promises to never use.

Nitrites are critical to the production of bacon, cold cuts, sausages, and hot dogs since these foods are susceptible to the most dangerous variety of food poisoning, botulism. This potentially deadly affliction is caused by botulin, a protein that attaches to nerve endings and blocks the release of the neurotransmitter acetylcholine. Lethargy, vertigo, double vision, speech problems, and difficulty swallowing may appear within eighteen to thirty-six hours of exposure. Eventually, paralysis of the diaphragm and chest muscles can cause death.

The term "botulin," deriving from the Latin *botulus*, for sausage, was coined in the eighteenth century in Germany when a number of deaths were associated with sausage consumption. Although it wasn't known at the time, spores of a bacterium, eventually named *Clostridium botulinum*, are found in soil and transform into viable bacteria in a low-oxygen environment, cranking out botulin, the most toxic substance in the world. Just 0.03 micrograms are enough cause death! That is about a millionth of a grain of salt, ten times less than the lethal dose of cobra venom.

The inside of a sausage is a low-oxygen, or anaerobic, environment, conducive to the germination of bacterial spores, should these be present. That is a real possibility with meat since animal feed can harbor spores originating in soil residues. Spores can survive boiling, but the toxin they produce cannot. However, cold cuts are not heated, and although bacon and hot dogs are, they are usually not heated long enough to destroy the botulin that may be present. Nitrite inhibits the growth of the botulin-producing bacteria, virtually eliminating the risk of botulism.

Unfortunately, nitrite can also react with naturally occurring amines in food and in the body to produce carcinogenic nitrosamines. This risk can be reduced with the addition of sodium erythorbate, a molecule with a molecular structure and properties virtually identical to vitamin C, but it's even more effective at preventing the formation of nitrosamines. Furthermore, it is a potent antioxidant and speeds up the curing process. But Maple Leaf does not allow it because it does not occur in nature and has to be produced by fermentation of rice. It does allow the less effective "lemon juice concentrate" because it is "natural." This is pure folly. The safety and efficacy of a substance does not depend on is origin.

In another bizarre twist, sodium lactate is on Maple Leaf's no-no list but "lactic acid starter culture" is allowed. Sodium lactate is a preservative that reduces the risk of listeriosis. It is commercially produced from glucose by fermentation using a lactic acid starter culture. For Maple Leaf, it is apparently fine to produce sodium lactate within a food, but adding it is not acceptable.

The whole concept of food being somehow safer because it contains simple, pronounceable ingredients is flawed. What matters is the overall composition of a food in terms of fat, carbohydrates, protein, vitamins, minerals, and perhaps phytochemicals, substances found in plants that may have desirable health properties. So, what happens when you compare the nutritional label on a package of the "simple" Maple Leaf hot dogs to a standard variety? They are essentially identical. Both have 7 grams of fat, 5 grams of protein, 3 grams of carbohydrates, and 330 milligrams of sodium. Both are fine for occasional consumption. Wonder if the kids in the Maple Leaf spelling bee can spell *occasional*?

ABOUT THAT FRENCH PARADOX

The fermentation of carbohydrates by yeast to yield alcohol may be the oldest example of applied chemistry, even predating the ancient process of making soap from animal fat and wood ashes. It seems that historically, drinking trumped cleaning. Much has been written over the years about the pros and cons of alcohol, with virtually everyone agreeing that overindulging leads to problems ranging from impaired driving to liver disease. The much-discussed question, however, is whether moderate drinking, usually understood to be about one drink a day, is detrimental, beneficial, or innocuous.

The possibility that alcohol, red wine in particular, may be beneficial was raised in the 1980s based on the "French paradox." How could it be that the French, who ate buttery croissants, fatty cheeses, and cholesterol-laden foie gras, have an unusually low rate of cardiovascular disease and obesity? Could the explanation possibly be their higher consumption of red wine? Indeed, some studies do suggest that people who consume no alcohol are less healthy than those who drink moderately. This observation may, however, be less meaningful than it appears due to what has been termed the *sick quitter effect*. Some, who in surveys show up as non-drinkers, may have given up drinking because of some ailment and that, rather than abstaining from alcohol, may account for their poorer health.

Still, there is the possibility that some component of wine, other than alcohol, offers a protective effect. Wine contains literally hundreds of compounds, including many polyphenols, which are of interest because of their antioxidant property. One of these, resveratrol, has attracted a great deal of interest from researchers because mice live longer when it is incorporated into their diet. However, the amounts fed to the animals were far greater than what people could possibly consume from wine. Incidentally, peanuts, soy, grapes, and Itadori tea, which has long been used in Japan and China as an herbal remedy for heart disease and stroke, also contain the compound.

A study of 783 seniors in Tuscany, where red wine is abundantly consumed, casts a further cloud on the resveratrol theory. Researchers used mass spectrometry to analyze twenty-four-hour urine samples for resveratrol and its metabolites and found that people with the highest concentration of resveratrol metabolites lived no longer than those who had none of these compounds in their urine. There was no association of the metabolites with inflammatory markers, cardiovascular disease, or cancer rates. So, if it isn't alcohol, and if it isn't resveratrol, what then can account for the French paradox?

Some argue that the question is irrelevant because there is no paradox. Death certificates in France are more ambiguous than in other Western countries and many cardiac deaths are not recorded as such. Others propose various reasons why cardiovascular disease may indeed be less frequent in France. Portions are smaller, so the French eat less food. They eat more vegetables, they cook more meals at home, which means less processed food. Meals are social occasions, not gulped down on the run. The French consume less sugar, drink fewer soft drinks, don't eat snacks often, nap more, and take longer vacations. If there really is a French paradox, it is likely multifactorial.

Now getting back to alcohol. It is classified by the International Agency for Research on Cancer as a Group 1 carcinogen, meaning that it is known to cause cancer in humans. Alcohol is linked to cancers of the breast, liver, rectum, and the digestive tract. There is controversy about whether there is a threshold effect, that is, a dose below which alcohol causes no harm. The Million Women Study in the U.K. monitored the health status of over a million women for more than seven years and found that the risk of cancer increases with alcohol consumption, even if that is only one drink a day. Type of alcohol apparently makes no difference. The researchers concluded that when it comes to cancer risk, no amount of alcohol is safe. That seems to be corroborated by a fascinating study in the U.K. that compared a group of moderate to heavy drinkers who gave up all alcohol for a month with a similar group who did not alter their drinking habits. Giving up alcohol for a month significantly reduced blood pressure, weight, cholesterol, proteins associated with cancer, as well as insulin resistance.

Just because the negatives are stacking up when it comes to alcohol, doesn't mean that we should become a society of teetotalers. After all, there is more to life than evaluating everything in terms of risk. We ride roller coasters, we ski, we bicycle, we barbecue, we eat oysters, we engage in sex, all of which have some risk. The same can go for alcohol. I'll drink to that. But not much.

SUGAR'S EFFECTS ARE NOT SO SWEET

Sugar stands in the dock accused of crimes against humanity. It is not the naturally occurring sugar in food that is being charged, it is "added sugar" that is being prosecuted for its role in causing heart disease, obesity, diabetes, and possibly cancer. But is there enough evidence to convict it? The case against sugar is not ironclad, but it is very compelling, based on animal studies, established biochemistry, and human epidemiological studies. There are researchers who maintain that the problem isn't sugar per se, but the extra calories it adds to the diet. Others claim that sugar is toxic and its effects on health are due to its biochemistry and have nothing to do with extra calories. What everyone agrees on, is that we are consuming way too much of the sweet stuff.

A typical epidemiological study, such as the National Health and Nutrition Examination Survey (NHANES) in the U.S., asks a large number of people to fill out questionnaires about their dietary habits and then monitors their health status over a period of time. When sugar consumption was examined, the researchers found that people whose diet derived 10 to 25 percent of its total calories from added sugar were 30 percent more likely to die from cardiovascular disease than people whose intake was under 10 percent of total calories during a follow-up period of fifteen years. And those consumers whose diet contained more than 25 percent of calories from added sugar were almost three times more likely to die. Furthermore, when the relation of diabetes around the world to obesity and sugar consumption is examined, it turns out that it is too much sugar that is the culprit, not obesity.

The problem may well be the fructose component of sugar. Sucrose, or table sugar, is composed of glucose and fructose linked together. During digestion, sucrose breaks down into its components, and glucose is absorbed into cells where it provides energy. But fructose is processed by the liver with some being converted into glycogen, a storage form of carbohydrates that can provide energy when needed. Excess fructose, however, is converted into fat, which then leads to fatty liver and a condition known as insulin resistance, which is linked to obesity, heart disease, and diabetes. A further concern is that fructose interferes with the formation of leptin, the hormone that sends the satiety message to the brain. Animal studies corroborate the damaging effects of fructose, and it is also noteworthy that populations from countries that consume high-fructose corn syrup have higher rates of type 2 diabetes than populations that consume table sugar.

The World Health Organization now recommends that added sugar, that is sugar not naturally part of food, should not make up more than 5 percent of the total calories in our diet. That translates to about six teaspoons a day. We have a way to go, given that Canadians now consume a whopping twenty-six teaspoons a day! That of course is an average: teenage boys wolf down some forty-one teaspoons, while senior women only consume about twenty. Where is all that sugar coming from? A can of sugar-sweetened soft drink has about ten teaspoons, the same as an equivalent amount of "no sugar added" fruit juice. A smoothie can harbor more than twenty teaspoons, a serving of Froot Loops about eleven (that's 100 times more than Shredded Wheat), a candy bar around seven, and a donut, four. Then there is the hidden sugar, like four teaspoons in a serving of tomato soup and half a teaspoon in a slice of bread.

It isn't hard to see that the sugar adds up. But so what? What's wrong with sugar? After all, it's natural isn't it? And natural substances are better for us than those chemically concocted sweeteners, aren't they? Actually, no. Sugar is a problem. And that has nothing to do with whether it is natural or not. It has to do with what it can do as it cruises through our body. Weight gain is an obvious possibility. Extra calories translate to extra weight, and sugar can deliver a lot of extra calories. A hundred and sixty in a can of pop. You would have to run at a pace of five miles per hour for fifteen minutes to burn that off.

In everyday language, the term "sugar," normally refers to sucrose, the white crystals isolated from sugar cane or sugar beets. But to a chemist, "sugar" can mean any of a number of simple carbohydrates that have a sweet taste. We have already seen that sucrose is actually composed of two sugars, glucose and fructose, joined together. Lactose, the naturally occurring sugar in milk, is made of glucose and galactose. Upon digestion, these are broken down into their components, which then enter the bloodstream. Starch, a carbohydrate composed of many glucose units linked together, is also a source of glucose upon digestion. When it comes to weight gain, the source of the sugars doesn't much matter. Carbohydrates, be they starch or simple sugars, are a problem.

Now, for the first time ever, a national regulatory agency is poised to tackle the problem. An expert committee that advises the Swedish government has recommended that new guidelines focus on a low-carbohydrate diet as the most effective method for weight loss. This is a huge turnaround given that the scientific community has largely dismissed low-carbohydrate diets as fads. However, after taking two years to scrutinize some 16,000 published studies, the committee concluded that low-carbohydrate diets work, and that, surprisingly, in spite of being high in fat, such diets have no negative effects on blood cholesterol. Keep in mind, though, that the Swedish recommendation was for weight loss, not overall health. As mentioned earlier, whole grains and fruits are important components of a healthy diet.

As far as weight loss goes, it seems that we may have been barking up the wrong tree with our calorie-counting, low-fat schemes. Diet gurus like Dr. Robert Atkins, whom we dismissed as cranks, were on the right track. It turns out that the oft-repeated dogma that weight is totally determined by calories in and calories out is theoretically sound but is of little practical significance. That's because the effective calories available from a food are not equal to the calorie content as determined by conventional experimental methods. In other words, consuming 100 calories worth of fat is not the same as 100 calories worth of carbohydrate. Fats and carbohydrates go through different metabolic pathways with different energy requirements. They also have different effects on insulin, the hormone that to a large extent determines the ratio of carbohydrates and fats the body uses for fuel. A reduced-carbohydrate diet forces the body to burn its fat stores for energy instead of glucose, the usual prime source.

But the issue isn't only about weight gain. Obesity is of course a major problem, associated with diabetes, heart disease, and even cancer, but sugar seems to be a problem even aside from its link to obesity. A major study published in 2014 in the *Journal of the American Medical Association* found a clear link between added sugar intake and cardiovascular disease mortality even in the absence of obesity. Soft drinks specifically were linked to heart disease. Of course, an association by itself cannot prove that sugar is the culprit, but it is suggestive, especially when one takes into account that fructose, which is released when sucrose is digested, has been implicated in causing metabolic problems.

The WHO's recommendation of added sugar making up no more than 5 percent of total calories is an extreme challenge to a population now consuming about 15 percent of total calories as sugar. And it is a bitter pill for the sugar industry to swallow because such a cutback could translate to billions of dollars in lost revenue. So we will undoubtedly hear the usual arguments about moderation and how sugar can be part of a balanced diet. Well, that depends on how one determines what amounts to a balanced diet. The WHO's experts have stated that in their view a diet isn't balanced if more than 10 percent of calories come from sugar.

When making dietary recommendations, one always has to consider any potential downside. With curbing sugar intake, there isn't one. Sugar is not a dietary requirement. Of course, cutting down is hard because it tastes so good. And it is also hard to know where it hides. It may be listed as barley malt, evaporated cane juice, corn sweetener, maltodextrin, brown rice syrup, molasses, dextrose, glucose, and of course high-fructose corn syrup. Time to be on the lookout for all these. One easy way to cut down is to just drink water instead of pop. Life may not be quite as sweet, but it may well be longer.

TAXING SUGAR

A swig of cola tastes just right with a slice of pizza and a smoked meat sandwich goes down nicely with a black cherry soda. I'll admit to an occasional such indulgence. But for many, having a soft drink is not a rare treat. It's a common daily habit. And it is not without consequence. The consumption of sugar-sweetened beverages has been associated with an increased risk of obesity, cavities, diabetes, and heart disease. Although the evidence isn't conclusive (it rarely is in science), many researchers are convinced enough about the potential harm of excessive sweetened beverage consumption to call for a special tax on such products to discourage sales.

Obviously, the soft drink and sugar industries are deadly opposed to such a tax and have marshaled their forces to wage war against any such legislation, vigorously attacking any study that suggests a link between health problems and sweetened beverages. The industry sponsors its own studies, 85 percent of which find soft drinks to be harmless. On the other hand, 85 percent of independently funded studies find a link between such beverages and poor health. The usual message from the industry is that there is no risk to health from any food or beverage as long as these are consumed in moderation. More or less that is true, but soft drink consumption in North America, close to half a liter per person per day, is not moderate.

Civil libertarians, who maintain that government has no right to interfere with what people choose to guzzle, are also opposed to any tax on sugary beverages. Surprisingly, even some nutritionists object to the idea. They say that people are so fixed on their sugar consumption that if the price of soft drinks goes up, they will just cut down on other more nutritious foods such as fruits and vegetables. We certainly would not want to see people maintain their soft drink consumption at the expense of apples. On the other hand, we do have evidence from the tax on cigarettes that taxation can have an effect on health. Statistics show that every 10 percent increase in price of cigarettes reduces sales by 3 percent, translating to thousands of lives saved.

In 2014, Chile, the largest per capita soft drink consuming country in the world, with 75 percent of its population overweight or obese, said enough is enough, and introduced a tax on all drinks containing more than 6.25 grams of added sugar per 100 milliliters. That meant Coca-Cola, with 35 grams of sugar in a 330 milliliter can, would be taxed. Seems fair, given that this is some 10 grams more than what the World Health Organization considers to be ideal consumption for a whole day!

The tax worked. Purchases of soft drinks in Chile fell by 20 percent the first year! Mexico, second in per capita consumption, also introduced a tax in 2014, although a different form. All sugary drinks were taxed one peso per liter (roughly seven cents), resulting in a 5.5 percent drop in purchases the first year, and 9.7 percent the second year. More recently, in 2018, the U.K. introduced a tax equivalent to thirty cents per liter on all sugar-sweetened drinks containing 5 to 8 grams per 100 milliliters, and forty cents a liter on those having more than 8 grams per 100 milliliters, with the money going to fund physical activity programs in primary schools. It is too early to tell what impact the tax will have on consumption, but about 50 percent of soft drink manufacturers have already reduced sugar levels to avoid the tax.

Soft drinks do not bring anything to the table except of course taste. They have no redeeming nutritional value. In fact, they have a negative value. Studies show that, unlike snacks, consuming sugary soft drinks between meals does not have an effect on the calories consumed at the next meal. Nibble on nuts or even potato chips, and in general you compensate by eating somewhat less at the next meal. But soft drinks don't have that effect, so they just add to the total calories consumed, thereby contributing to the weight-control problems we are witnessing.

If more convincing is needed to cut down on soft drinks, how about looking at the results of a 2019 study by researchers at Harvard's T.H. Chan School of Public Health? They followed over 100,000 men and women who had regularly filled out questionnaires about their diet and health status and found that over the thirty or so years of the study, the risk of death was associated with the consumption of sugar-sweetened beverages. Men who drank at least two sodas a day had a 29 percent higher risk of death when compared with those who drank less than one a month while women had a 63 percent higher risk. Overall, subjects who drank two or more sodas a day were 31 percent more likely to die of cardiovascular disease and 18 percent more likely to die of cancer. Even those who drank artificially sweetened beverages were at greater risk of death, but that may have been because people with health problems such as obesity or diabetes switched to diet drinks.

When it comes to taxation, or perhaps a red dot warning as Canada is contemplating, why pick on soft drinks and not on other sweet concoctions such as chocolates? Several reasons. While chocolates are generally recognized as a treat, soft drinks are consumed on a regular basis. Furthermore, a typical drink has 35 to 40 grams of sugar, which according to many experts is roughly the maximum amount that should be consumed per day! That's enough that it should be a very occasional indulgence.

SUGAR AND THE BRAIN

The bitter news about sugar just keeps coming. Especially when it comes to fructose. Most people recognize fructose as the sugar found in fruits, where it creates no problems. But fructose also makes up half of the sucrose molecule and when sucrose is digested, it breaks down into fructose and glucose, and it is the fructose part that researchers say is particularly problematic. They've linked it with obesity, diabetes, and fatty liver. And now, some preliminary data in rats suggests that fructose can even impair brain function. The media has latched on to the study carried out at UCLA's School of Medicine with headlines implying that a sugary diet can make you stupid. If you are a rat.

In the experiment, rats were first taught to navigate a maze, a standard requirement if you're a laboratory rodent. They were then divided into two groups, and both groups were given drinking water laced with fructose for six weeks. But one group was also given a mix of omega-3 fats because these have a reputation for improving brain function. After six weeks, the rats were challenged with the maze once more and the animals that had not been fed the omega-3 fats performed more poorly. Basically, they were much worse at remembering the route through the maze that they had learned six weeks earlier. For some reason the researchers did not have a control group of rats that was fed normal laboratory food. It would have been interesting to see how they fared compared with rats on the high-fructose diet.

In any case, the high-fructose rats also showed insulin resistance, which was likely responsible for the impaired memory. Insulin is involved in regulating how brain cells use and store sugar for the energy required to process thoughts. The researchers' conclusion is that omega-3 fats as found in flaxseed, walnuts, and fish can protect the brain from the harmful effects of fructose. But let's not make too much of a rat study. Let's remember that there was plenty of stupidity around before we started to guzzle sugar in such stunning amounts. Still, stupidity does seem to be increasing. And more and more people are complaining of impaired memory. Maybe they can't remember that they are being urged to consume less sugar.

While sugar can be considered to be a dietary villain, it has also been accused of a crime of which it is innocent. Contrary to the perception of many parents and educators, sugary snacks do not make children bounce of the walls.

A British television program, *The Truth About Food*, decided to put the sugar-hyperactivity link to a more or less scientific test. The producers organized two parties for children. When parents dropped their kids off for the first one, they saw tables loaded with sugary snacks. But as soon as they left, the junk food was replaced by healthy snacks and the kids were entertained with some high energy music. Two weeks later, the same children were invited to a party, this time with a sedate storyteller providing the entertainment. A feast of healthy snacks was laid out for the parents to see, but it was quickly replaced by cakes, cookies, and soft drinks after they left.

After each party, parents were asked to evaluate their children's behavior, and there was consensus that the first party had made them hyperactive. The parents were not surprised, having seen the sugary snacks to which their kids had been supposedly treated. Had the ruse not been revealed, they would have reinforced their conviction that sugar cause hyperactivity. In truth, the ill behavior was caused by the excitement of the party, the frantic music, the running around. The second party was a calm affair, and the children were delivered to their parents in a peaceful state despite the high sugar load.

A small study organized by a television program may be interesting, but it isn't peer-reviewed science. However, there have been peer-reviewed publications that have examined the link between sugar consumption and hyperactivity. A 1995 meta-analysis of such studies published in the *Journal of the American Medical Association* found that sugar does not affect the behavior or cognitive performance of children.

Just because there is no scientific support for sugar causing hyperactivity, doesn't mean we let it off the hook as a nutritional culprit. Flooding the bloodstream with sugar causes a burst of insulin to be released, which then can drop the sugar level in the blood quickly, sometimes to below normal levels. This may result in muddied thinking and poor classroom performance. But other nutritional factors, such as the type of fats in the diet, likely play a bigger role in determining children's behavior.

Fats are an integral component of cell membranes and determine the fluidity of these membranes, which in turn affects the way cells communicate with each other through chemicals called neurotransmitters. With the advent of processed foods, our pattern of fat consumption has changed. Trans fats, byproducts of hydrogenation, and omega-6 fats found in corn and soy oil have increased at the expense of omega-3 fats, found in fish and vegetables. This can conceivably affect behavior, and some studies have indeed shown an improvement in children's behavior when the diet is supplemented with omega-3 fats. There is also some evidence that gluten in wheat or casein in milk can in some cases have an adverse effect on behavior, as can certain food additives such as MSG or tartrazine, a yellow dye. While these factors are debatable, there is no doubt that a diet with fewer processed foods and less sugar is preferable for all sorts of reasons. So certainly give kids apples and carrot sticks instead of cakes and ice cream at the next party, but if you want good behavior, hire a cellist instead of a clown.

HOW SPLENDID IS SPLENDA?

They tortured that poor molecule. They heated it, froze it, dissolved it in acid, baked it into cakes, stuffed it into people's mouths and fed it to rodents. They even made it radioactive so they could follow its path through the animals' bodies. Then they extracted it from the rodents' poop to see how it had fared. And it fared well. In fact, it passed every indignity with flying colors. As a result, sucralose (Splenda) was introduced in 1991 in Canada and has since become a leader in the artificial sweetener sweepstakes.

We love sweets. There's no doubt about that. Our palates lust for ice cream, our mouths water at the thought of glazed donuts, our parched throats yearn for soft drinks, and visions of sugar-plums sometimes dance in our heads. To satisfy these cravings, we gorge on caloric sweeteners such as sugar or high-fructose corn syrup. This is not exactly ideal for our teeth, our waist-line, or our general health. Thus, it isn't surprising that artificial sweeteners and artificially sweetened products have attracted greater and greater consumer interest.

Unfortunately, none of the sugar replacements have been completely satisfactory. Aspartame, or Equal as it is branded, is about 180 times sweeter than sugar but is not stable in acidic conditions or when exposed to heat. This presents significant problems for diet drink manufacturing and baking. Saccharin and the cyclamates are sweet enough but leave an aftertaste. There have also been some lingering health concerns. Aspartame cannot be consumed by individuals with an inherited condition called phenylketonurea (PKU) because their bodies lack the ability to metabolize phenylalanine, one of this sweetener's breakdown products. In rare cases, people may have adverse reactions to aspartame, including headaches and visual disturbances. Saccharin and the cyclamates, in turn, have had to cope with the shadow of cancer. Some studies, probably of no relevance to humans, have suggested a slightly elevated risk when test animals were fed massive amounts of these sweeteners. In any case, the nutritional world was primed for a new kid on the block.

The kid was born in a laboratory at Queen Elizabeth College, University of London, in 1976. The researchers studying the chemical reactions of ordinary table sugar, or sucrose, certainly did not have artificial sweeteners on their minds. But when they managed to incorporate three chlorine atoms into a sucrose molecule, they aroused a sugar company's interest. A company representative called one of the researchers to ask for a sample to be tested. As luck would have it, the young foreign chemist misunderstood and thought the request was for a sample to be "tasted." So with a bit of bravado, he plopped some of the chlorinated sugar into his mouth and told his supervisor about the sweet experience. The supervisor immediately recognized the potential value of this discovery, seized the moment, and redirected the laboratory's research efforts. Sucralose, as the new compound came to be called, turned out to be 600 to 1,000 times sweeter than sugar, depending on what it was added to!

But many years of testing faced the energized chemists before the new product could be brought to market. Their enthusiasm increased when sucralose turned out to be very water soluble as well as stable to heat and acid. This meant that it could easily be used in diet drinks and baked goods. Since sucralose is so sweet, much less of it is needed than if sugar is used. But this presents a problem. Sugar provides not only sweetness but bulk in bakery products. However, when sucralose is combined with maltodextrin, a type of starch that provides bulk, the mixture can be substituted for sugar, measure for measure. The end product isn't necessarily identical, however. Since sugar is also responsible for the browning effect produced by baking, the color of some cookies, for example, may look rather anemic. One thing sucralose is not good for is making fudge. It comes out much too syrupy.

Is there anything, then, to criticize about sucralose? We might be wary of some of the marketing approaches that have trumpeted the safety of this compound by referring to the fact that it is made from "natural sugar." What a substance is made from is irrelevant; what matters is the final product. Its properties are determined not by its ancestry, but by its molecular structure. Hydrogen gas, for example, can be made from water, but it would be absurd to suggest that it therefore has the same safety profile. It's a different substance, just like sucralose is different from sugar. Incorporation of three chlorine atoms into the sugar molecule converts it into a totally new substance. Sucralose is safe because it has been extensively tested, not because it is made from sugar. One of its other attributes is that it leaves no bitter aftertaste, but unfortunately, I cannot say the same thing for some of the advertising hype about the product.

The battle for conquest of the sweetness receptors on our taste buds is a vicious one. It is being fought in supermarket aisles, in restaurants, and increasingly, in courtrooms. The brutal conflict pits the sugar industry, and its "natural sweetener," against artificial sweetener interests and their low-calorie products. At stake are billions of dollars in profits, and according to some, the health of the public. Both sides sponsor umbrella organizations to further their cause. The Sugar Association's mission is to "promote the consumption of sugar as part of a healthy diet and lifestyle through the use of sound science and research." The Calorie Control Council claims essentially the same mandate, but of course on behalf of the diet product industry. There are some curious bedfellows here, since low-calorie sweeteners also tangle with each other for market share, with producers of sucralose and aspartame commonly sniping at each other. For the public, it's an unhappy situation. But the lawyers reap profits.

Those of you who have followed my writings, lectures, or media presentations over the years will know that I am no great fan either of artificial sweeteners or sugar-laden foods. I am a fan of good science, and I resent the cherry-picking of data and the use of misleading terminology to further any cause. And there is plenty of both in the sweetener wars. An example would be the intense battle between the sugar industry and manufacturers of sucralose, commonly sold as Splenda.

Ever since its introduction in 2000, sucralose has been taking a bite out of sugar profits and has managed to relegate aspartame, the artificial sweetener that once dominated the market, to second place. Capitalizing on the fact that the raw material for the production of sucralose is sugar, Splenda built an advertising campaign around the phrase "made from sugar, so it tastes like sugar." Both the sugar and aspartame producers took issue with this slogan and launched lawsuits claiming the advertising was misleading.

There are two problems with Splenda's original slogan. First, "so it tastes like sugar" does not logically follow from "made from sugar." Once the molecular composition of sugar has been altered, the properties of the new material do not necessarily bear any similarity to the properties of the starting material. It would be ludicrous to say, for example, that since hypochlorous acid can be made by replacing an atom of hydrogen in water with one of chlorine, its properties are those of water. Second, "made from sugar," suggests a closer connection with sugar than actually exists. Since many people, albeit wrongly, believe that natural products are always safer than synthetics, an association with "natural" sugar is useful for marketing purposes. Sucralose has an excellent safety profile, but that has nothing to do with its being made from sugar. It has to do with the extensive testing that was carried out before approval and with the scarcity of adverse reactions that have been reported.

Although the manufacturer of Splenda maintains that its advertising is honest, the company did agree to an out-of-court settlement with Merisant, the company that markets aspartame under the trade name Equal. The battle with the Sugar Association, on the other hand, continues. And it is a dirty one. An Association sponsored website, the Truth about Splenda, aims to present sucralose in a most unfavorable light and invites consumers to submit their comments about the sweetener. Surprise, surprise, the comments are all bitter. They are, however, worth reading, because they do offer an insight into the chillingly poor state of scientific knowledge among the public.

Let me take an example. A consumer writes, "I was appalled that a product sold as made from sugar is actually made from chlorine. Would you put bleach into your coffee? Well, using Splenda, that is actually what you are doing." Actually, you are doing nothing of the sort. True, both bleach and Splenda contain chlorine atoms, but they are completely different substances. The use of chlorine in the production of sucralose has no bearing on the safety of the product.

Another correspondent is "heartsick to find out that Splenda contains chlorine." Would he be also heartsick to find out that the hydrochloric acid in his stomach, critical for digestion, also contains chlorine? One more. "I never used sugar substitutes because I didn't want to put chemicals in my body. I started to use Splenda because it was made from sugar. Now I find out it contains chlorine. This false advertising needs to stop." Well, if you don't want to put chemicals into your body, you'll be dining on a vacuum. It isn't very nutritious. And what is the false advertising? Not disclosing the molecular composition of Splenda? Nonsense. Asking that chlorine be listed as an ingredient, as another writer demands, is absurd. As absurd as asking that hydrogen be listed as an ingredient if water is present.

While Splenda's advertising can be justifiably criticized, how about the Sugar Association's own tagline? "Sugar: sweet by nature." Doesn't that imply that sugar is safe because it is natural? Botulin is made by nature, but I sure wouldn't want to eat it. Sugar is safe not because it is natural, but because research and epidemiological evidence have shown it to be so. More or less.

It isn't particularly safe for diabetics, and it plays a role in tooth decay. There is even suspicion that fructose, one of the components of sugar, may be linked with alarming changes in body fat and insulin sensitivity.

When sucralose was originally introduced it had already undergone extensive testing for some fifteen years. Short-term and long-term animal-feeding studies had shown that most of a sucralose dose was excreted unchanged, and even the small percentage that was metabolized yielded compounds that were also excreted. Every bit the animals were fed could be accounted for in their excreta. Any concerns about storage in the body or interference with metabolic pathways essentially evaporated. As an added benefit, unlike sugar, sucralose was shown to have no detrimental effect on the teeth. While our bodies cannot break down sucralose, microorganisms in water and the soil readily do so. In other words, the stuff is biodegradable and poses no environmental hazard.

Because of the widespread use of artificial sweeteners, researchers continue to have an interest both in their effectiveness and safety. A number of studies have shown that using artificial sweeteners does in most cases not lead to weight loss, possibly because they increase the preference for sweet tasting food. Another possibility is that sweeteners change the composition of gut bacteria, favoring microbes that play a greater role in the breakdown of food components leading to more efficient absorption instead of excretion. Israeli researchers have in fact shown that artificial sweeteners can alter the composition of gut bacteria. Another issue has cropped up with the use of sucralose in baking. With prolonged heating it decomposes to yield hydrochloric acid as one of the breakdown products. This can then react with glycerol, a fat decomposition product, to form chloropropanols, which in sufficiently high doses can be toxic. There is no evidence, however, that the tiny amounts formed from sucralose pose a risk to people.

While there appears to be no major concern with artificial sweeteners, there is no great benefit either. Instead of replacing a sugary beverage with a diet drink, why not replace it with water? And while that slice of cake may be yummy, so is a bowl of berries.

TOM BRADY'S ALKALINE TWADDLE

He has never eaten a strawberry. In fact, he avoids all fruits except for the occasional banana. No sugar, no white flour, not even a sip of coffee pass his lips. He doesn't eat tomatoes or green peppers or any other nightshade vegetable. Everything he eats must be organic and conform to an "alkaline" diet. He is Tom Brady, one of the most successful quarterbacks in football history!

What Tom, however, is not, is an expert in nutrition. Neither is his "lifestyle guru," Alex Guerrero, whose academic background is a degree in traditional Chinese medicine from a defunct institution. He has been fined by the Federal Trade Commission in the U.S. for presenting himself as a doctor and claiming to cure a variety of diseases ranging from cancer to AIDS with the dietary supplement Super Greens.

That didn't stop him from giving poppycock another go with Neurosafe, a beverage that supposedly protected against "sports-related traumatic brain injury" with its content of creatine, magnesium, zinc, and ethylenediamine tetraacetic acid (EDTA). There is no evidence that any of these ingredients help in recovery from head trauma, but nevertheless Tom Brady endorsed the product: "Neurosafe makes me feel comfortable in that if I get a concussion, I can recover faster and more fully." The FTC wasn't as comfortable and confronted Guerrero, resulting in Neurosafe being taken off the market.

As far as nutrition goes, Guererro's philosophy is revealed in his book, *In Balance for Life: Understanding & Maximizing Your Body's pH Factor*. This was basically a rehash of Robert O. Young's 2002 epic twaddle, *The pH Miracle*, the basic tenet of which was that an "acidic" body is susceptible to disease and eating alkaline foods not only protects against illness but is curative. Where did Young's scientific expertise come from? Well, he does have degrees from the Clayton College of Natural Health, an unaccredited diploma mill that has since ceased operation!

There is no scientific validity to disease being caused by an acidic pH, and in any case, blood is a "buffered" system, meaning that it resists any change to its normal pH of about 7.35. Neither diet nor supplements can alter this to any significant extent, and to suggest that "alkalizing" can cure cancer is criminal. Young eventually found that out the hard way.

In 2014, he was charged with practicing medicine without a license, along with various other crimes, including theft. The alkalizing guru was accused of steering patients away from conventional care and charging high fees for his own treatments. He was convicted of practicing medicine without a license (the jury was divided concerning theft), sentenced to three and a half years in prison, and had to openly declare that he had no degrees from any accredited institution and was not a scientist of any kind. He was also sued by Dawn Kali, a former breast cancer patient who claimed that Young had presented himself as a medical doctor and advised her to forego surgery or chemotherapy in favor of "pH injections." In 2018, a jury awarded Kali a stunning US$105 million, finding that Young had been negligent and his treatments fraudulent. Justice done.

While Tom Brady's adherence to an alkaline diet is based on faulty theory, the diet itself may have some benefits in that it features lots of vegetables and limits red meat, processed foods, sugar, soft drinks, and alcohol. Eliminating fruits, though, is misguided. But what about Tom's worship of organic foods? That would seem to be supported by a recent study in France that had close to 70,000 adults fill out questionnaires about their diets, specifically asking whether sixteen specific foods were consumed in their organic or conventional versions. They were then followed for up to four and a half years and asked to report any diagnosis of cancer.

Publication of the study triggered a media frenzy with headlines screaming that eating organic foods can reduce the risk of developing cancer by 25 percent! While that is sort of true, in science, the devil is often in the details. After correcting for confounding factors such as physical activity, family history, socioeconomic status, processed food intake, and red meat consumption, the incidence of cancer in the quartile of the subjects that consumed the fewest organic foods was 2.2 percent, while that in the highest quartile group was 1.7 percent. The difference between these is about 25 percent. More realistically, according to this study at least, for every 1,000 people eating organically, five cases of cancer may be prevented. That still sounds pretty good!

Close examination of the data, however, reveals, that levels of significance were only reached for women and only for postmenopausal breast cancer and non-Hodgkin's lymphoma. Furthermore, perhaps the key finding, one that was skimmed over in the media accounts, was that among participants who had a diet of high overall quality, namely eating lots of fruits and vegetables, there was no difference in cancer rates! The real takeaway message from this study then is to eat your fruits and vegetables, whether they be conventional or organic. And Tom, that includes strawberries and nightshade vegetables. Passing on these has no impact on your passing.

SILLY JILLY

I don't get angry easily. Yes, I get annoyed when I see homeopaths promoting absurd "nosodes" as alternatives to vaccines, chiropractors "adjusting" babies' spines, televangelists hawking "Miracle Spring Water" for protection from disease, and scientifically illiterate bloggers claiming that any food ingredient with an unpronounceable name should not be consumed. My annoyance, however, has recently switched to anger in two particular cases in which farcical health advice is being offered by self-proclaimed experts.

Brittany Auerbach, who goes by the name Montreal Healthy Girl on social media, presents herself as a "naturopathic doctor." This is based on having taken an online course from the unaccredited Institut de Formation Naturopathique (IFN), whose president claims to have a degree from the nonexistent Lincoln College of Naturopathic Physicians and Surgeons. Students who take the IFN's online course never see a patient and cannot be licensed as naturopaths. They are of course still free to produce videos and pontificate on all sorts of subjects in which they have no expertise, as Auerbach routinely does.

The video that really ticked me off was one in which she claimed that "cancer is actually a good thing," because it is a warning that the body is too acidic, a condition that will result in death unless an alkaline diet is followed. She maintains that "all diseases are reversible with the proper lifestyle changes and positive outlook," that HIV and Ebola infections can be treated with colloidal silver, and refers to chemotherapy and radiation as "hideously ludicrous."

Physicians, Auerbach argues, induce panic in cancer patients because of their ignorance and lack of understanding about cancer. She, on the other hand, is "a certified naturopathic doctor [really?] who has done extensive research on cancer" and knows that "making your body ideal for healthy cells will cause all cancer cells to die over time." This drivel is found in a letter to her own uncle after he had been diagnosed with throat cancer. Instead of undergoing chemotherapy, which according to an oncologist had a chance of beating the disease, the man chose to treat himself with wheatgrass juice and an alkaline diet. It didn't work. One wonders how many others have been seduced by such apparent simple solutions to a complex problem.

Brittany Auerbach may be scientifically confused, but she still takes a back seat to Jillian MaiThi Epperly, who soars to new heights with her stunning ignorance and outlandish claims. I had not heard of this spectacularly wacky woman until I was alerted to her appearance on the *Dr. Phil* show. Never before have I heard such concentrated hogwash in such a short time. Epperly claims that the cure for every disease from AIDS to cancer is over-salted fermented cabbage juice that rids the body of all disease-causing toxins and parasites by triggering "waterfall" diarrhea. Never mind that the high salt concentration can trigger dangerously high blood pressure. That's not all. Drinking "Jilly Juice" will allow us to extend our life expectancy to 400 years! It even "cures" homosexuality, and get this, regrows missing limbs! Epperly proudly admits she has no relevant education, which according to her is a plus because it is educated doctors who have failed to curb all the diseases that afflict us.

I think there are two possibilities here. Epperly is either missing a few screws, or she has come up with a devilish scheme to get attention that translates to monetary gains by making claims that are more outlandish than the most absurd lunacies that permeate the Internet. Indeed, Jillian has managed to slither onto the set of *Dr. Phil* and undoubtedly procure some new paid subscriptions to her website, where she rambles on about her miraculous juice. The host did appropriately discredit Epperly, who would need a fair bit of education to rise to the level of being uneducated, but still, giving her exposure will cause some desperate people to follow her insane advice.

Neither of these women is spreading benign poppycock. They both steer people away from effective treatments, spread false hope, and assail science. Auerbach says that if she were diagnosed with cancer she would not even think of conventional therapy. Epperly claims to have a protocol "to reverse 100 percent of all health issues" and advocates giving Jilly Juice to babies. I claim that both are full of baloney. Or bunkum. Or balderdash. Take your choice. I have other words too.

There is some good news, though. At the McGill Office for Science and Society we have managed to bring the antics of Montreal Healthy Girl to the attention of the media, triggering television and radio investigations that resulted in her taking down the asinine cancer videos. I am also pleased to see that there are a number of groups in the U.S. that are trying to stop Epperly from spreading her mental muck on social media and that the FDA is looking into her activities. Unfortunately, it is hard to take legal action because she is not selling the juice. She is only selling claptrap, which unfortunately is legal. Perhaps now you understand my anger.

THE DIRT ON CLEAN EATING

I thought I was a clean eater. I wash my fruits and vegetables. I scrupulously scrub my cutting board after use. If I thaw meat in the fridge, I make sure the juices do not contact any other food. I don't buy into the "five-second rule," so I don't eat food that has dropped on the floor. I'm also careful to keep my meals from attacking my shirts and ties. But it seems that according to a number of "wellness" books that are flooding the market and numerous memes on social media, I'm not really "eating clean."

Why not? Because I'm not worried about eating gluten, I don't seek out organic produce, I'm not reviled by preservatives, I have not forsaken eggplant, tomatoes, or other nightshade vegetables, and I don't think GMOs present a health issue. I do not avoid all dairy, I eat the occasional croissant, and now and then, I even indulge in a smoked meat sandwich. Only at Schwartz's of course. Furthermore, I do not worship at any altar dedicated to kale, beetroot, coconut oil, or Himalayan salt. My spiralizer does not have a permanent place on my counter, and I do not think coconut sugar is any better than regular sugar. I don't think bone broth has magical properties, I think that alkaline diets are bunk, and I do not judge the safety of a food ingredient by the complexity of its chemical name. In the eyes of the current promoters of "clean eating," all of that would make me a dirty eater.

The concept of "clean eating" emerged within the last decade as a path towards a healthier life. It is based on the belief that the modern "Western" diet of fatty, salty, sugar-laden, and highly processed foods might as well be the scythe wielded by the Grim Reaper. There is no simple definition of what "clean" means, but avoiding various foods deemed to be unhealthy, as well as all processed foods, is a central theme. Gwyneth Paltrow, that icon of nutritional wisdom, eats clean by shunning sugar, alcohol, dairy, soy, corn, and nightshade vegetables. Her preferred health beverage is cultured goat milk. And to make her skin "glow" she drinks Goopglow Superpowder that contains vitamins C and E, CoQ10, lutein, zeaxanthin, and grape-seed anthocyanins. The claim is that this goop fights free radicals, provides photo protection, and helps build collagen. There is zero evidence that this product improves the appearance of the skin.

Gwyneth's Goop Inc. also sells Psychic Vampire Repellent Protection Mist that among its ingredients lists a complex blend of gem elixirs, sound waves, moonlight, love, and reiki charged crystals. One would think that to repel vampires, sunlight would be more appropriate than moonlight. And strangely, there is no garlic in the concoction. Goop also promotes the work of Anthony William, the "Medical Medium" who passes on medical advice that he gets from a spirit, as well as that of Caroline Myss, a "medical intuitive" whose gift of "biospiritual consciousness" allows her to "sense illness and disease in people before they've been diagnosed by a physician." Plenty of gumption but no evidence.

Neither is there any evidence that the vitamin B12 injections that Gwyneth's devotees line up for at the Goop Health Summit conference "offer mental clarity and energy, while clearing brain fog." Too bad because they seem to be suffering from mental obscurity and brain fog. And should someone suffer from fatigue, they can avail themselves of Goop's Why Am I So Effing Tired, described as "a comprehensive vitamin and dietary supplement regimen designed to promote adrenal function, mental acuity, and stress tolerance." Effing ridiculous.

Jasmine and Melissa Hemsley, British food writers and media personalities with a huge following, push astrologically farmed vegetables and eggs, whatever that may mean, along with bone broth and agave nectar. No dairy, starch, or gluten for these ladies, and no carbohydrates eaten together with proteins. Then there is Jordan Younger, a "wellness" blogger who rose to fame as the "Blonde Vegan," advocating a raw, organic vegan diet that is free of gluten, sugar, oil, grains, and legumes. Her "5-day cleanse" program, featuring an assortment of green juices and based on the absurd notion that this would rid the body of a host of unnamed toxins, snared thousands.

It seems, though, that Jordan's dietary regimen was not entirely conducive to wellness given that her periods stopped, her hair began to fall out, and she constantly felt exhausted. On a friend's advice, she finally added salmon to her diet, which made her feel better. She felt obligated to disclose this to her fans, and that elicited a host of vile comments from her dedicated vegan followers for transitioning away from veganism. Some demanded their money-back for her cleanse programs and accused her of not having the discipline to be truly clean. There were even threats on her life.

Jordan's journey down the clean eating road was prompted by digestive problems that at first seemed to be alleviated by eliminating certain foods. But then her approach became more and more extreme until she was in the throes of orthorexia, a term coined by Dr. Steven Bratman in 1997 from the Greek words for "right" and "appetite." Orthorexia is an obsession for eating only what are perceived to be healthy foods and eliminating "unhealthy" ones. It is an eating disorder that can compromise health.

Since food is the only raw material that enters our body, we are quite literally built of its components. It therefore makes sense that diet and health are linked. That notion is not new, and historically "impure" foods have been pilloried, although the view of what constitutes impure food has changed. While in the 1800s the concern was about "greening" pickles with copper compounds or preserving meat with arsenic, today's dietary boogeymen are GMOs, gluten, sugar, and various chemicals in processed foods. There is certainly nothing wrong with a quest for a healthy diet, as long as it is based on science. That cannot be said for most of the "clean" dietary regimens promoted by an assortment of bloggers and celebrities with no relevant scientific background who beguile the public with their recommendations to stay clear of "chemicals" and engage in detox schemes.

I try to consume mostly whole grains, lots of fruits and vegetables, some nuts, not much red meat, few fried foods, little salt, and no soft drinks. That's clean enough for me.

MEDICAL MEDIUM

Celery growers are thrilled. The vegetable is selling like hot cakes. There have been runs on celery before, usually coinciding with the appearance of some article claiming that more calories are needed to digest this vegetable than it provides. While it is true that celery contains only about 6 calories per stalk, there is no food that results in "negative calories." But that bit of nonsense is nothing compared with the truckload of detritus that is being dumped on consumers by promoters of celery juice who claim that it is a veritable cure-all. The "brains" behind this puffery, and the man responsible for the boost in celery sales is Anthony William, who calls himself the "Medical Medium."

William freely admits that he has no medical education of any kind, but then again, he doesn't need any. That's because he gets his information from an all-knowing spirit who somewhat unimaginatively is actually called Spirit and speaks to William in a clear voice that only he can hear. Let's let William tell the story himself, as he does in his book, *Medical Medium*, that, believe it or not, made the *New York Times* bestseller list.

> My story begins when I'm four years old. As I'm waking up one Sunday morning, I hear an elderly man speaking. His voice is just outside my right ear. It is very clear. He says "I am the Spirit of the Most High. There is no spirit above me but God." Later that evening, I suddenly see a strange man standing behind my grandmother. He has gray hair and a gray beard and is wearing a brown robe. When none of my family reacts to his presence, I slowly realize that I'm the only one who sees him. He says, "I am here for you." Then the gray man looks at me, "now say, Grandma has lung cancer."

William then recounts how his grandmother was shaken by this bit of information coming from a four-year-old, and even though she felt fine, she made an appointment for a general checkup. A chest x-ray revealed that she had lung cancer! And so with the help of a man that only he sees, speaking with a voice that only he hears, at the young age of four, a career as a medium with a talent for diagnosing and treating disease was forged.

When you open *Medical Medium*, subtitled *Secrets Behind Chronic and Mystery Illness and How to Finally Heal*, to chapter one, you can feast your eyes on the first sentence: "In this book, I reveal truths you won't learn anywhere else. You won't hear them from your doctor, read them in other books, or find them on the web." I think that is true. According to William, the reason we don't hear of the "truths" elsewhere is because Spirit has "insights into health that are decades ahead of what is known by medical communities."

Spirit also teaches William to perform diagnostic body scans. His training though is not in hospitals but in cemeteries. "I spent years in different cemeteries performing this exercise with hundreds of corpses. I became so good at it that I can almost instantly sense if someone's died of a heart attack, stroke, cancer, liver disease, car accident, suicide, or murder." Quite a gift. Although it seems the technique may need a bit of work, judging by William giving an "all clear" diagnosis to a TV host who was soon after diagnosed with malignant myeloma. William doesn't need to be in the same room with a person to do a reading. He can do it by phone, as he demonstrates on his radio show. He asks callers for their symptoms, and then diagnoses them, usually as suffering from an infection by the Epstein–Barr virus, and prescribes a treatment, which often is 16 ounces of celery juice. Some would call this practicing medicine without a license.

There is no such thing as an autoimmune disease, William says, and multiple sclerosis is a version of the Epstein–Barr virus that can be cured with a host of supplements such as barley juice extract powder. According to William, attention deficit hyperactivity disorder (ADHD) as well as Lyme Disease (an infectious disease caused by bacteria that William absurdly claims is a viral infection) can also be treated with supplements, and digestive problems respond to, guess what, celery juice. Indigestion, we are told, is a problem because the six components of hydrochloric acid in the stomach (a ludicrous notion) don't work well together. Celery juice works its magic because it contains "unique sodium compositions." Spirit seems to be in need of a chemistry lesson.

William recognizes that there are problems with plastics, but Spirit has a solution for this as well. "Anti-Plastic Tea" is a blend of equal parts of fenugreek, mullein leaf, olive leaf, and lemon balm. This is different from "Anti-Radiation Tea," a blend of Atlantic kelp, Atlantic dulse, dandelion leaf, and nettle leaf. William can also diagnose, with Spirit's help of course, Alzheimer's disease. When a lady was exhibiting memory problems, William scanned her and found "two large pockets of mercury in the left hemisphere of her brain." A heavy metal detox regimen with barley grass juice extract powder, spirulina (preferably from Hawaii), cilantro, wild blueberries (only from Maine), and Atlantic dulse was the prescription.

Incidentally, Spirit is also a talented car mechanic, advising William on fixing "unfixable" cars to the utter shock of legitimate car mechanics. And oh yes, William also sees angels. One helped him save his drowning dog. There are twenty-one essential angels, he tells us, and they can be called upon for help, but the call must be out loud and the specific angel has to be named. You can call upon the Angel of Purpose if you are struggling with your purpose on Earth, and you can ask the Angel of Water to change the frequency of the water you bathe in to make it more cleansing, nourishing, and grounding. There are also exactly 144,000 "Unknown Angels" who don't have names but "are eager for the chance to work with us, and by summoning them we can tap into a resource of profound power for healing the body, mind, heart, spirit, and soul."

And if the angels aren't up to the task, you can go for the birds and the bees. "Birds sing the song of the angels and the heavens and can mend fractured souls and reverse disease." How does that happen? "The frequency of these melodies resonates deep within our DNA, which allows it to reconstruct the body on a cell level." If you don't care to listen to birds, you can watch bees. "As bees dance from flower to flower, they emit a healing frequency that reverses disease and promotes soul and emotional restoration."

If you are not keen on listening to birds or watching the bees and are not comfortable calling out loud to angels, there's always celery juice. There is nothing special about this juice. Like any vegetable, celery does contain a host of chemicals, notably apigenin and luteolin, that have antioxidant, anti-inflammatory, and anticancer properties. However, these effects have only been observed in rodents and cell cultures using the isolated compounds, not the vegetable or its juice. There are hundreds of compounds found in fruits and vegetables that have such properties which is why the best bet is to eat a wide selection. There is no single "superfood" or beverage.

Celery juice does have one practical use. It is a source of nitrates and nitrites and is now used in processed meats so they can declare "no added nitrate or nitrite" on the label. Nitrates and nitrites are a concern because of potential conversion in the body into carcinogenic nitrosamines. Of course, the source of nitrite is irrelevant. Whether it comes from celery juice or from a bottle purchased from a chemical company makes no difference. For a 150-pound person, 16 ounces of celery juice provides roughly three times the daily recommended dose of these chemicals. So drinking all that celery juice is not such a great idea. Apparently Spirit can also use a course in food science. Maybe in economics as well.

Depending on the price of celery, which does vary seasonally, it can cost from four to six dollars a day to produce the amount of juice that William recommends. If you don't want to squeeze it yourself, it is available commercially for around seventeen dollars for 200 milliliters. That works out to about forty dollars per day if you follow William's protocol. Or you can buy celery juice powder for $165 for 500 grams. The packaging says, "Contents may do you good." Or they may not.

What can we conclude about the Medical Medium? There are only three possibilities. One, spirits exist, and there is at least one with medical knowledge that is superior to that possessed by any physician and that it has selected Anthony William to impart this knowledge to ailing people. Two, William is a clever huckster who capitalizes on people's desperation. Three, William is a candidate for psychiatric care. You decide.

SPARE ME THE WHEATGRASS ENZYMES

"Just try it once, please, just try it," the lady begged me. "Okay," I finally said, hoping to bring the discussion to an end. She opened the thermos bottle she had been clutching and poured me a glass of a green liquid, assuring me that she had squeezed the wheatgrass barely an hour ago. I could therefore be confident, she said, that the enzymes in it were still alive! Well, dead or alive, they certainly did nothing for the taste of the beverage. This wheatgrass juice was one of the foulest things I've ever tasted. Of course, I was quickly assured that I was not drinking it for taste; I was drinking it for health.

This gustatory calamity followed on the heels of an hour-or-so-long discussion on the merits of consuming chlorophyll and live enzymes. My guest had sought an appointment to open my eyes to a form of therapy that would help millions of people who were being poisoned by eating "dead food." And so it was that I came to learn about the Hippocrates Health Institute and the teachings of Ann Wigmore.

Ann was a Lithuanian émigré to the U.S. who had become convinced of the healing power of grasses after reading the Biblical story of Nebuchadnezzar, the Babylonian king who apparently cured himself of a seven-year period of insanity by eating grass. Wigmore reflected on this story, considered how dogs and cats sometimes eat grass when they feel ill and came up with a theory about the magical properties of wheatgrass juice.

Food rots in the intestine due to improper digestion, she maintained, and forms toxins that then enter the circulatory system. The living enzymes in raw wheatgrass prevent these toxins from forming and ward off disease. By 1988 Wigmore, who had no recognized scientific education, was even suggesting that her "energy enzyme soup" was capable of curing AIDS.

Ann Wigmore died in 1994, but the live enzyme theory lives on. Numerous books tout the benefits of ingesting enzymes, health food stores stock bottles of enzyme capsules and powders, and restaurants that guarantee low-temperature cooking to prevent the murder of enzymes are sprouting up. No need to worry about killing enzymes, though. They were never alive in the first place. Enzymes are not composed of cellular units; they cannot reproduce, they cannot carry on metabolism, and they cannot grow. Ergo, they are not alive.

"There would be no life without enzymes," begins the usual sales pitch. Well, you can't argue with that statement. Indeed, enzymes are special protein molecules that are involved as catalysts in virtually every chemical reaction that takes place in the body. "Heat can destroy enzymes," the pitch continues, "so processed or cooked food is devoid of these life-giving substances." This is also true. The inference then is that we should be eating "live food" because that's the only way we can get the "live enzymes" our body needs. In the case of Ann Wigmore, it is more than an inference. Her book states: "Each of us is given a limited supply of enzyme energy at birth. This has to last a lifetime. The faster you use up the supply, the shorter your life. Cooking food, processing it with chemicals, using medicines, uses up the enzymes. The Hippocrates Diet makes enzyme deposits into the account." Absurd! Our body does not need ingested enzymes, and, except for specific rare instances, it cannot use them.

Enzymes are proteins, and, like other proteins, they are broken down during digestion. The fact that studies have shown that some enzymes may escape digestion and enter the bloodstream should not be interpreted as a benefit. Enzymes are remarkably specific in their actions, and the enzymes that may make it into the bloodstream from food are not the same as the body's enzymes. Many promoters of live food diets emphasize that the "living enzymes" in fresh fruits and vegetables help digestion and spare the body's enzyme supply from being wasted on digestion. The spared enzymes are then said to be free to take part in metabolism and disease fighting. Nonsense. Metabolic enzymes have nothing to do with digestive enzymes. Even if enzymes in raw fruits and vegetables survived passage through the highly acidic environment of the stomach and managed to enhance digestion in the small intestine, they would have no effect on the enzymes involved in the cellular processes that go on all over the body.

This is not to say that oral enzyme therapy is always without merit. People who are lactose intolerant can benefit from ingestion of the enzyme lactase, which is lacking in their digestive tract. But the lactase pills have to be specially formulated to enhance passage through the stomach. Cystic fibrosis patients have to compensate for a lack of pancreatic enzymes by swallowing pills, which are enterically coated to ensure they reach the small intestine. In some cases, people with digestive problems may benefit from plant-derived enzyme supplements that help break down proteins, fats, and carbohydrates. While these enzymes do the job in the laboratory, their effectiveness in the digestive tract is controversial. Researchers are also investigating whether certain oral enzymes may be of use in cancer treatment, but unfortunately, so far, the results haven't been particularly encouraging.

There is yet another bizarre feature of the raw food diet

championed by Ann Wigmore and others. They point out that the molecular structure of chlorophyll, the green coloring in plants, is almost the same as hemoglobin, which carries oxygen around the body. They infer that ingesting chlorophyll enhances energy by increasing oxygen transport. This is pure twaddle. Humans are not plants; we do not photosynthesize, and we have no requirement for chlorophyll. In any case, chlorophyll cannot be absorbed.

Now that I've gotten all that off my chest, I'll go on record as recommending a live food diet. The fruits and vegetables that make up such a diet contain all sorts of substances that enhance health. But enzymes are not among them. As I related all of this to my office guest, I had a glimmer of hope when she accepted my explanation that oral enzymes in food are unlikely to survive digestion. The apparent victory, though, was short-lived. "Maybe that's why Ann Wigmore was so high on wheatgrass juice enemas," she retorted. Mercifully, she didn't ask me to try one.

RAW WATER NONSENSE

There's dumb, dumber, superdumb, and hyperdumb. Then there is dumb that is beyond a suitable qualifier. Promoting "raw water" as a healthy commodity is in that category. This is water that has not been treated in any way. No filtration, no chlorination, no ozonation, no ultraviolet light treatment, in other words, none of those nasty technologies that save millions and millions of lives a year.

Depending on the source, raw water may be perfectly safe, or it may lead to a battle with hepatitis, giardiasis, or a host of bacterial diseases. You could get away with nothing more than cramps, vomiting, or explosive diarrhea, but you could also end up in a close and personal relationship with raw underground water. Six feet underground.

Promoting untreated water in the face of everything we know today about waterborne disease is so preposterous that it shouldn't even merit discussion. But the raw water craze can't be ignored because people are wading into the claptrap. Where? Mostly in California, the very state that introduced Proposition 65, requiring warnings like "this Disneyland resort contains chemicals known to the state of California to cause cancer and birth defects or other reproductive harm" because there might be a rogue vacuum cleaner somewhere with a touch of lead solder. No real risk with that, but apparently California has no worries when it comes to the real risk of untreated water.

Live Water is a brand of raw water for which some folks happily fork out $36.99 for two and a half gallons. It is the brainchild of Christopher Sanborn, a man who rechristened himself as Mukhande Singh, perhaps because that name goes along better with pictures of this longhaired raw water guru sitting naked in a yoga position, apparently floating above some remote spring.

As he tells the story, one day Singh had a revelation. He discovered that "all bottled water is filtered, sterilized, and irradiated for cheaper transport and shelf stability similar to how juice and milk products are pasteurized to save costs." Whoa! These measures are not carried out to save costs, but to save lives. He then goes on to say that "unfortunately this destroys five healthy probiotic strains not found in any other food source, and without these probiotics we are unable to assimilate all the nutrients found in our food."

"Probiotic" is a popular catchword these days, and for good reason, given that there is accumulating evidence of the role that gut bacteria play in our health. Probiotics are "good bacteria" that are thought to stave off potentially disease-causing varieties. Surprisingly, Singh is clever enough to capitalize on this idea. As long as nobody asks for any evidence.

Because of his concerns about the travesties inflicted on bottled or tap water, Singh set out on a search for water untainted by the hand of man and found salvation in a spring somewhere in Oregon. "The first time I drank living spring water, a surge of energy and peacefulness entered my being; I could never go back to drinking dead water again." And apparently, he knows all about dead water. It's "toilet water with birth control drugs in them," he proclaims, the quality of his grammar being comparable to the quality of his science.

By contrast, Singh's Live Water contains a host of probiotics, or at least bacteria he calls probiotics. Analysis of Live Water does reveal the presence of various bacteria such as *Pseudomonas putida*, one of his supposed probiotics. There is no evidence that this bacterium confers any benefit, but I can point out a paper in the medical literature with the title "A Lethal Case of *Pseudomonas putida* Bacteremia Due to Soft Tissue Infection." By mentioning this, I don't mean to imply that *Pseudomonas* bacteria make Live Water dangerous; just that there is no evidence they make it "healthy."

Trying to put some oomph into the claim that "living spring water is the key to unlocking a perfect microbiome balance," Live Water promotions refer to a scientific publication entitled "Non-pathogenic Microflora of a Spring Water with Regenerative Properties." Alluring, until you take the trouble to read the paper. It has nothing to do with drinking spring water! The paper describes "experimental fresh wounds in an animal model showing reduced inflammation when treated with Italian Comano spring water." The authors hypothesize that this may be due to beneficial bacteria found in the water and suggest that these microbes may even explain why people believe that bathing in spring waters can benefit human skin ailments. Absolutely nothing to do with drinking Live Water!

Live Water does come from a remote spring, and there is no evidence that this particular raw water harbors disease-causing organisms. But what the whole concept of raw water does harbor is enough detritus to pollute science and possibly compromise health. Anyone for playing Russian roulette with raw water?

SOURED ON LEMON JUICE

"Lemon tree, very pretty, and the lemon flower is sweet, but the fruit of the poor lemon is impossible to eat." The lyrics of the classic Peter, Paul, and Mary song ring true enough. We don't peel lemons and eat them. Far too sour thanks to their content of citric and malic acids. Take lemon juice, though, mix with some sugar water, and you get very pleasant tasting lemonade. Some people forego the dilution and scoff down pure lemon juice with hopes of health benefits. Numerous bloggers scream about drinking lemon juice to boost immunity, fight cancer, prevent kidney stones, or to just "detox." No validity here, but lemon juice has played an important role in disease prevention historically, as long as the disease we are talking about is scurvy.

Up to the end of the eighteenth century, seafarers suffered from the scourge of scurvy, a terrible disease in which collagen, a main structural protein in the body, cannot be replaced. This leads to the breakdown of tissues with symptoms ranging from bleeding gums and sores to shortness of breath and potentially fatal heart problems. Today, scurvy is known to be caused by a deficiency in vitamin C, but that wasn't firmly established until Hungarian physician and researcher Albert Szent-Györgyi isolated the compound from paprika in 1932. It was given the name ascorbic acid, from the term "antiscorbutic," used to describe substances that prevented scurvy. Such substances, mostly fruits and some vegetables, had been identified centuries earlier by some explorers, but their findings did not get spread around extensively. Jacques Cartier, for example, learned from natives that a brew made from spruce needles could treat scurvy and Captain James Cook provided sauerkraut to his crew with good results. However, it was the seminal work of Scottish physician James Lind in 1747 aboard the Royal Navy's ship, the HMS *Salisbury*, that laid the foundation for the treatment of scurvy.

Lind knew that scurvy was rare on land but common on long ocean voyages. Like others before him, he postulated that diet might be involved and put his idea to a test in what was a pioneering controlled clinical trial. Twelve sailors with symptoms of scurvy were divided into six pairs, with each pair being treated with remedies that had been previously suggested, ranging from cider and dilute sulfuric acid to garlic and citrus fruits. Within a week, the men who had been given two oranges and a lemon every day were well while the others' condition worsened. Lind published his *Treatise of the Scurvy* in 1753, but back in those days such publications didn't circulate as efficiently as today, and it took forty-two years before the British navy began to distribute lemon juice to sailors on a daily basis.

Interestingly enough, there were still problems. Sometimes the lemon juice was heated to preserve it, but heat destroys vitamin C. Also, when lemon juice is stored for a long time, vitamin C breaks down. Lind had proposed that the power of the juice lay in its acidity, and that led to the replacement of lemons by limes, which were just as acidic but were more readily available from British colonies than lemons. This led to British sailors being referred to as limeys. Unfortunately, limes only have half as much vitamin C as lemons, so this replacement led to an upsurge in scurvy. An exception were sailors who broached the problem of a monotonous diet aboard ships by hunting rats and eating them. Rats, unlike humans, can synthesize their own vitamin C and do not have a requirement for this vitamin in their diet. Sailors were protected as long as the rats were not overcooked.

As is often the case, when a substance shows a wondrous effect against some medical condition, it is tried for others as well. Perhaps the best publicized investigations of vitamin C focused on its potential use in the treatment or prevention of the common cold. This use was championed by Linus Pauling, double Nobel Prize winner and one of the greatest scientists of the twentieth century. Pauling's evidence was anecdotal, but because of his fame, people began to pop vitamin C to ward off colds. Then scientists stepped into the game and decided that a proper investigation was in order. Since the 1970s some fifty-five placebo-controlled studies which used at least 200 milligrams of vitamin C supplements a day have examined the link between vitamin C and the common cold, with ambivalent results. It seems people who take vitamin C regularly suffer no fewer colds but may experience a slight reduction in the number of days with symptoms. Marathon runners, skiers, and soldiers who are exposed to significant cold and or physical stress are an exception, actually showing a reduction in the incidence of colds.

As far as lemons go, don't expect them to add zest to your life. But to your salad, for sure!

MATCHA AND MACA

Social media is abuzz with chatter about matcha tea and maca root. Since I'm not adverse to expanding my dietary horizons, I was game to engage in a little exploration.

Unlike regular green tea made by extracting the components of tea leaves with hot water, "matcha" tea is prepared by stirring powdered green tea leaves into water, then whisking it with a brush to produce a foamy, grassy-tasting beverage. Not any old tea leaves, mind you, but ones that during the final stages of growth have been shaded from the sun with black netting before being stone-ground into a powder. Lack of sun forces the leaves to produce more of the green pigment chlorophyll, as well as more caffeine, epigallocatechin gallate (EGCG), and theanine, meaning that drinking a matcha concoction is sort of like drinking concentrated tea. The espresso of teas, as it were. Matcha has a long history of use by the Japanese, especially Zen monks, in traditional tea ceremonies, but recently it has become a hot item here, promoted as some sort of super healthy beverage.

EGCG is a member of the class of compounds known as polyphenols, which have antioxidant properties. Although there is a theoretical basis for suggesting that antioxidant intake can mitigate disease and slow aging, there is no compelling clinical evidence. What is known is that consuming plant-based products seems to confer benefit, perhaps due to polyphenols, but of course plants contain numerous other compounds that may play a role. Theanine is an amino acid that is said to have a stress-relieving or anti-anxiety effect, but proper studies are lacking.

Drinking matcha means that the whole leaf is consumed, so that more EGCG, theanine, and caffeine are available than from a regular tea infusion. Some studies have suggested that EGCG may even have a role in weight control since it increases metabolism, meaning that more calories are "burned." But there are no studies that link matcha with weight loss. Sometimes matcha powder is incorporated into pancakes, ice cream, lattes, or cheesecake to take advantage of its superfood aura. This is just marketing gimmickry that ignores the large sugar content of these foods. Drinking unsweetened matcha, however, was a surprisingly pleasant experience.

Now on to maca. A good story can sell a product, especially when it comes to dietary supplements. Talk about some legendary use by natives, throw in terms like "increased stamina," "improved mood," "aphrodisiac," and "natural," and you are off and running to the marketplace. Maca is grown mostly in Peru, and its root, with a composition much like wheat or rice, has a long history as a dietary staple. But it is stories about the enhanced virility of Inca warriors who supposedly downed maca root before going into battle that captured the imagination of supplement manufacturers.

Couple this with anecdotes of Peruvians eating maca root for energy and improved sexual function, and you have a basis for carrying out studies that may potentially lay the groundwork for sound science. After all, plants are fascinating chemical factories, and it is conceivable that maca may have some biologically active compounds. None have been detected so far, but that is not surprising. It takes a monumental effort to isolate, separate, and identify the hundreds of compounds found in plants, and that is only the beginning. Then comes the even greater challenge of testing candidate compounds for biological activity. That's why when it comes to herbal products, the simplest process is to test crude mixtures.

In one small study, men taking 1,500 or 3,000 milligrams per day of powdered root claimed increased sexual desire compared with a placebo. There was no measurable change in sex hormones and curiously the effect was not dose dependent. Another study in young men showed a slight but significant improvement in erectile dysfunction, and one in postmenopausal women resulted in decreased anxiety and depression with some improvement in sexual function compared with placebo. Again, there were no changes to any hormone levels.

As usual with such dietary supplements, the consumer is at the mercy of the manufacturer in terms of product quality. Regulators do not systematically check that the product actually contains what it is supposed to contain. However, given that maca is widely consumed as a food, it is unlikely that any of the root powders pose a significant health risk, although headaches, stomach problems, sweating, and sleep disruption have been reported in rare cases. It seems that for people looking for a little boost in stamina and sexual function, a daily dose in the range of 1,500 to 3,000 milligrams of Peruvian ginseng, as maca is sometimes called, is an option. It may actually do something, especially if you think it will.

I've been told I should mix some maca into my matcha for a match made in heaven. Hmmmm . . .

THE SAGA OF KOSHER COKE

"With the help of God, I have been able to uncover a pragmatic solution," declared Tobias Geffen, the chief rabbi of Atlanta, in 1935. What was the problem in need of a solution? Whether or not Coca-Cola was kosher!

The famous beverage had been introduced as a health tonic by Atlanta pharmacist John Pemberton, who had been wounded in the Civil War and had become addicted to morphine, the only substance that controlled his pain. Concerned about his addiction, Pemberton searched for an opium-free painkiller and came up with an alcoholic concoction laced with extracts of the coca leaf and the kola nut. The combination of alcohol with cocaine from the coca and caffeine certainly would have dulled the pain. When Atlanta introduced temperance legislation in 1886, Pemberton had to remove alcohol from his tonic. He looked for novel ways to make it palatable. He added sugar along with a mix of fruit juices and plant extracts, keeping the exact composition a secret as protection from would-be competitors. Legend has it that one day he accidentally used carbonated water to formulate his product and was so taken by the taste that he decided to market "Coca-Cola" as a fountain drink.

Pemberton's health was poor, and in 1888, he sold the company to another Atlanta pharmacist, Asa Candler, who later sold it to a group of investors headed by Robert Woodruff. Under Woodruff, the company grew into a global concern with clever marketing highlighting the secret nature of the formula, supposedly known only to two executives who were never allowed to travel together.

By the 1930s, the popularity of the beverage was enticing Jewish consumers who wondered if it conformed to the complicated laws of kashrut. Four-legged animals destined to be eaten have to chew their cud, have split hooves, and have to be butchered in a certain fashion. Meat and dairy cannot be mixed, insects are forbidden, and only fish with scales can be consumed. The production of kosher food has to be supervised by a mashgiach, an inspector who is trained in the ways of kashrut. Contrary to what many people think, the dietary laws were formulated based on spiritual, not bodily health. The idea was that adherence to kashrut would form a unifying bond among Jews and would serve as a constant reminder of the importance of a relationship with God, who had laid down the dietary laws.

When it came to Coca-Cola, there was a problem. The company had built its marketing campaign largely around the secret ingredients in its formula, and getting approval as being a kosher product would mean having to disclose the ingredients for evaluation. There was much hesitation about this, but finally the company decided that it would be financially too painful to lose the Jewish market and in 1935 made a deal with Rabbi Geffen allowing him to scrutinize the formula as long as he promised to reveal only information pertaining to kashrut.

The Rabbi found two problems. Glycerin, used as a sweetener and preservative, was made from nonkosher beef fat. Since glycerin can also be produced from plant products, Coke's supplier, Procter and Gamble, easily replaced animal fat with cottonseed and coconut oils. A second concern was that some of the flavors used in Coke were extracted with alcohol, and since the source of the alcohol was not clear, there was the possibility that it had been fermented from grains that were not kosher for Passover. This was solved by using alcohol that had been fermented from sugar found in sugar beets or sugar cane. Once these changes were made, Coca-Cola received approval as a kosher beverage. Observant Jews rejoiced, at least until Coke replaced cane sugar with high-fructose corn syrup as its sweetener.

During Passover, corn and products made from it are avoided by some segments of the Jewish population, notably the ones originating in Eastern Europe, the Ashkenazi Jews. Back in the thirteenth century, rabbis worried that some people might confuse flour made from corn with flour made from wheat, barley, rye, oats, or spelt, grains forbidden during Passover. They decided that the best way to prevent this was by also banning any food, such as corn, that could be confused with the banned grains. To ensure sales during Passover, Coca-Cola now produces a "kosher for Passover" version of its classic beverage that is made with cane sugar instead of high-fructose corn syrup. Diet Coke contains no corn products, so there is no issue with it being suitable for Passover.

Rabbi Geffen did indeed uncover a pragmatic solution. Whether it was with the help of God or not will remain a mystery. But it was certainly with the help of chemists.

FCKD UP IS GONE. GOOD RIDDANCE.

On March 1, 2018, the body of teenager Athena Gervais was discovered face down in a creek in the Montreal suburb of Laval. The coroner's report concluded that she had stumbled into the creek after becoming intoxicated. Athena had purchased three "energy drinks" that contained 11.9 percent alcohol and consumed most of these within about twenty-three minutes, resulting in a blood alcohol level of 192 milligrams per 100 milliliters, way above 80 milligrams per 100 milliliters, which is the legal limit for driving in Quebec. The beverage she consumed, a product described as a "caffeinated, high-alcohol, high-sugar drink," had the repugnant name FCKD UP. It has since been removed from the market, and Quebec has lowered the legal limit of alcohol in beverages that can be sold in grocery stores to 7 percent.

When this story first emerged, one of my students asked me if it is okay to mix Red Bull with alcohol. Good question. Alcohol is a central nervous system depressant while caffeine is a stimulant. And, as Mary Poppins told us, "a spoonful of sugar makes the medicine go down." The net effect of caffeine and sugar is to mask the onset of drunkenness and increase consumption by creating an illusion of sobriety. Aside from the common problems associated with overindulgence in alcohol, there may be some extra risk when alcohol is combined with caffeine. Red Bull is a non-alcoholic energy drink that contains both sugar and caffeine, hence my student's valid question.

In Canada, the addition of caffeine to alcoholic beverages is illegal, but "natural flavors" can be added. This provides a loophole for squeezing caffeine into the drink in the form of guarana extract. Guarana is a climbing plant native to South America that produces a fruit with seeds containing about four times the amount of caffeine found in coffee beans. The seeds' effects were recorded as far back as the seventeenth century, when a Jesuit missionary noted that members of an Amazon tribe who consumed the seeds had "so much energy that when hunting, they could go from one day to the next without feeling hungry."

When it comes to including guarana in a drink, the only requirement is that its presence be identified on the label. There is no obligation to list the amount of caffeine, and depending on how the seeds are extracted, the content can vary greatly. After the teenager's unfortunate death, Health Canada tested FCKD UP and found the amount of caffeine to be negligible, meaning that it was not a combination of alcohol with caffeine that resulted in the teenager's death. Guarana, however, is not off the hook.

Plants are veritable chemical factories, capable of producing hundreds of compounds. It's possible components of guarana other than caffeine have a physiological effect. Researchers at Colgate University in the U.S. examined this possibility using the planarian, an aquatic flatworm, as a model system. These creatures have a central nervous system that uses neurotransmitters comparable to those found in mammals, so they are commonly used in studies of stimulants. Their friskiness, or locomotor activity, in response to exposure to specific substances can be readily observed and conclusions can be drawn about possible effects in humans.

Curiously, caffeine was not found to have a marked effect on planarian motility, but guarana extract provided stimulation, especially when combined with glucose. Specific chemicals responsible for the activity have not been identified, but the plant is known to contain theophylline and theobromine, which both have recognized biochemical activity. Furthermore, guarana contains tannins, compounds that can form complexes with other molecules, including caffeine, possibly leading to slower release and a longer-lasting stimulant effect. More research is needed, but evaluating the effects of the combination of alcohol with guarana may have to go beyond considering just a caffeine-alcohol reaction.

A caffeine-alcohol reaction, however, is the main concern when it comes to the question of mixing an energy drink such as Red Bull with an alcoholic beverage. Energy drinks promise to stimulate both the body and the mind with a combination of ingredients, which in the case of Red Bull amount to caffeine, taurine, and various B vitamins. Taurine is widely found in animal tissues and owes its name to the Latin term for "bull," since it was first isolated from the bile of that animal in 1827. While taurine has a number of biological functions, there is no evidence that it has any sort of energizing effect. It is likely added to justify the drink's name, which is intended to conjure up an image of becoming "strong as a bull." Neither do B vitamins produce any sort of stimulation, so caffeine is the "energizing" component of Red Bull.

The literature on energy drinks is extensive and reveals many cases of emergency room visits, mostly due to palpitations and arrhythmias, both when such beverages are consumed with alcohol or alone. A randomized controlled trial has even shown that energy drinks produced more electrocardiogram aberrations than a control beverage containing an equivalent amount of caffeine. This suggests that further evaluation of non-caffeine ingredients in energy drinks is warranted. In virtually all cases of adverse effects, though, consumption was high, over 500 milliliters.

Bottom line? Consuming one alcohol–Red Bull mix is unlikely to be a problem. More could be. And there should be more control over the way that alcoholic energy drinks are sold. The Quebec coroner has accordingly recommended that the federal government ban the sale of alcoholic energy drinks with names or images that downplay the problems of excessive consumption, drunkenness, or alcohol addiction. Hopefully governments will expend some energy in controlling the marketing of alcoholic energy beverages.

POPPY SEED TEA

It has probably caused more misery than any other drug in history. It has also caused more relief from misery than any other drug in history. It is morphine. It is also the drug with the oldest history of use by humans, going by the *Ebers Papyrus*, the earliest known medical treatise. Found in a mummy's tomb in Luxor, this document, written around 1550 BC, describes the dried latex of the poppy as a remedy for noisy children. It undoubtedly was that, since opium, as the latex came to be called, can induce sleep. Unfortunately, in the wrong dose, it can induce sleep permanently.

Managing the appropriate dosage of morphine became possible after German apothecary Friedrich Wilhelm Serturner isolated the compound from opium in 1805, earning him several honorary degrees from universities along with a cash prize as a "Benefactor of Humanity." Indeed, many a suffering patient has benefited from the pain-relieving properties of morphine. But, as is often the case in science, there is another side to the story. Use and abuse often travel parallel paths.

The majority of drug overdose fatalities in North America are due to opioid abuse, a consequence of addiction. A quest for the euphoria produced by morphine, its synthetic derivative heroin, or analogues such as oxycodone or fentanyl, is a prime cause of addiction. However, addiction can also occur when patients are prescribed opioids for legitimate reasons. A further emerging problem is the availability of unwashed, unprocessed, unregulated poppy seeds online for brewing into poppy seed tea by people looking for a high.

The morphine content of poppy seeds is variable, depending on the type of poppy and where it is grown. In general, seeds contain little morphine, with most of the plant's pharmacologically active compounds, which also include codeine, thebaine, and papaverine, being found in the poppy's sap. Still, when poppies are mechanically harvested, the sap can contaminate the seeds. Producers of seeds intended for food use rely on plants bred to contain lower levels of morphine than those cultivated for pharmaceutical use. Food seeds are also washed to further reduce any residual active compounds. Nobody is going to get high on a poppy seed bagel, muffin, or Hungarian sweet poppy noodles. I can vouch for the latter because it was a staple in our house when I was growing up. It's a simple dish made by tossing egg pasta with a bit of melted butter and lots of ground poppy seeds mixed with powdered sugar. Never gave me the slightest buzz. Did excite the taste buds though.

Nevertheless, I don't think I would want to indulge in that dish if I were involved in some sort of competition that required a urine test for opioids. While there isn't enough morphine to produce a pharmacological effect, there may just be enough to be detected by the super sophisticated analytical techniques now available to monitor substance abuse. That classic *Seinfeld* episode in which Elaine fails a drug test because of a poppy seed muffin is not pure fiction. Indeed, in Britain a worker was fired from his job at a power station after a routine drug test showed opiates in his urine that he maintained must have come from the poppy seed bread he had eaten for breakfast. His story came to the attention of *Rip Off Britain: Food*, a popular television program, and resulted in the host proving the point. Angela Rippon ate a loaf of poppy seed bread along with a poppy seed bagel over three days and ended up testing positive for opiates.

Since the poppy seeds used in bakery items are generally washed, they are unlikely to contain opiates in doses high enough to cause any notable effect other than possibly showing up in the urine. Poppy seed tea is a different matter. Some of the seeds sold online, or in bulk in stores, are not properly washed and can contain appreciable amounts of morphine. Teas brewed from these can reach potentially deadly levels. When Professor Madeleine Swortwood of Sam Houston State University was contacted by the parents of Stephen Hacala, a young man who had died after drinking poppy seed tea, she decided to investigate whether brewing tea from seeds bought from twenty-two different vendors could yield a beverage with potentially fatal results. Indeed, she was able to extract concentrations of morphine, codeine, and thebaine that could be lethal with moderate consumption.

After being made aware of these findings, Stephen Hacala's parents launched a campaign to make people aware of the danger of teas brewed from unwashed poppy seeds. There have been at least twelve deaths in the U.S. since 2010 attributed to poppy seed tea consumption, with most of the seeds being bought online. The Hacalas have met with FDA and Department of Justice officials, alerting them to the problem and urging that random sampling and testing of poppy seeds available online be carried out and that sales of unwashed seeds be stopped.

A bizarre case has shown that poppy seeds can kill even without producing toxic levels of opiates in the blood. A fifty-four-year-old woman with epilepsy had read on the Internet that eating poppy seeds would make her feel better. They didn't. They killed her. Cause of death was determined to be complications of a bowel obstruction secondary to massive poppy seed ingestion. Autopsy revealed a cast-like obstruction in the large bowel composed of poppy seeds.

Problems notwithstanding, this discussion rekindled my taste for poppy noodles. I made a batch but cut back on the sugar in the recipe. There's emerging evidence that it's addictive.

ACTIVATED NUTS

The question took me by surprise. "Do you really have to activate your nuts?" Just as a plethora of strange mental images began to bubble through my mind, I learned that we were talking about tree nuts. "Activation" does not refer to poking sleeping nuts; it refers to soaking and then drying raw varieties before consuming them. The aim is to increase nutritional value. And profit margins for processors. Although "activated" on the label is synonymous with "expensive," affluent health conscious consumers have made bags of activated nuts trendy. Are they activating anything other than cash registers?

The hype of "especially healthy" foods is rarely corroborated by science, but there is usually a kernel of truth that is germinated and then fertilized with cherry-picked data to yield a crop of questionable value. "Activated nuts" follow this pattern.

Nuts are the seeds of plants and are equipped with the nutrients needed for plant growth, including a host of proteins, fats, starches, minerals, vitamins, sterols, and protease inhibitors, all needed for a seed to germinate and go on to produce a plant. Sterols are essential components of plant membranes, and protease inhibitors are proteins produced by plants to protect themselves from insects by interfering with the action of proteases, enzymes insects use to digest plants they have dined on. Basically, they give insects a tummy ache. But protease inhibitors can also interfere with human digestion. Soaking leaches them out, reducing the problem.

One of the essential elements needed for plant growth is phosphorus, which seeds store as a component of phytic acid. However, phytic acid can also bind essential minerals such as iron, magnesium, calcium, and zinc that may be present in food, preventing them from being absorbed into the bloodstream during digestion. Phytic acid is also found in beans, grains, and various seeds, meaning that a high consumption of these can result in mineral deficiency. Soaking beans and nuts unleashes a host of chemical reactions that collectively are referred to as germination. Enzymes called phytases are generated to break down phytic acid and release phosphorus, as well as any bound minerals for easier absorption. All this suggests that "activation" by soaking beans, grains, and nuts improves their nutritional value. But there is another side to phytic acid and phytates, the term for compounds in which minerals are bound to phytic acid. They have purported health benefits, so there may be a consequence to eliminating them from the diet.

One of the risk factors for colon cancer is too high an iron intake, linked to high meat consumption. Vegetarians have a lower rate of colon cancer, and it's been suggested that they are protected by their intake of phytic acid, which reduces iron absorption. So, should we listen to the paleo community that tells us that phytic acid is an "anti-nutrient," and that foods containing it should be shunned, to the "nut activators" who claim that soaking nuts makes them healthier by eliminating phytic acid, or to researchers who have linked phytic acid to reduced risk of cancer as well as diabetes, kidney stone formation, and heart disease? What about supplement manufacturers who market inositol hexaphosphate, another term for phytic acid, with oblique suggestions of efficacy in the treatment of cancer, heart disease, depression, and side effects of chemotherapy based on some inconclusive preliminary experiments?

Each of these positions can be supported by selective reporting of data, but the bottom line is that neither phytic acid nor phytates are drugs or poisons, and for the vast majority of the public, their presence or absence in nuts is irrelevant. We do not rely on nuts as our sole source of minerals or of phytic acid. A balanced diet will provide adequate amounts of both. In any case, the nuts we consume are not truly "raw." They have been either roasted or blanched to inactivate natural toxins, and in the process, their phytic acid content has been significantly reduced, and soaking provides no further benefit.

Nuts, whether activated or not, can make a valuable contribution to the diet. A number of studies have shown a reduced incidence of heart disease and gallstone formation associated with nut consumption, as well as trends towards reduced risk of hypertension, cancer, and inflammation. Interventional studies have shown that nuts can have a cholesterol reducing effect, likely due to their sterol content. Sterols have a chemical similarity to cholesterol and can block cholesterol absorption. Despite their high fat content, epidemiologic studies and clinical trials suggest that nuts are unlikely to contribute to obesity and may even help in weight loss. Just about the only health issue is the possibility of an allergic reaction. And there is another bonus with nuts. They taste good. I buy the roasted, unsalted variety. They are active enough for me.

ARE LECTINS NUTRITIONAL CRIMINALS?

Lectins are a class of proteins found in many foods. They can kill you! All you have to do is extract some from castor beans, introduce them into a tiny perforated metal pellet, load this into the tip of a special umbrella fitted with a spring, aim the umbrella at the thigh of an intended victim, and trigger the spring to inject the pellet into his bloodstream. He'll be dead in a couple of days. That's not theory, that's the way a Bulgarian agent assassinated dissident Georgi Markov in London in 1978. Ricin, the lectin found in castor beans, is one of the most powerful poisons in existence. No question it can be lethal. If it is injected into the bloodstream.

Why bring up a forty-year-old murder? Because of a book that is terrifying readers with the claim that "lectins are the number one biggest danger in the American diet." To avoid messing up our internal machinery with these dastardly toxins, we have to curb our intake of a host of foods that include whole grains, bread, beans, corn, soy, tomatoes, peppers, nuts, eggplant, and all fruits, except when they are in season. If we do that, we will solve our digestive problems and eliminate worries about weight gain, high cholesterol, arthritis, and "brain fog."

Who says so? Dr. Steven Gundry, author of the bestselling *The Plant Paradox*, the paradox being that plant foods that are supposed to be good for us actually harbor hidden dangers. I first heard of Dr. Gundry when I was forwarded a link to his video that I painfully watched for close to an hour. It basically turned out to be a slick commercial for his "energy boosting" polyphenol-containing Vital Reds, his metabolism boosters, his probiotic blend, and his vitamin and mineral supplements. It was disenchanting to see a respected cardiac surgeon pushing supplements not backed by proper studies.

Without question, Dr. Gundry once was an accomplished cardiac surgeon with an arm-long list of credits, including the invention of devices used in heart surgery. Then in the early 2000s, as he explains, he discovered that lives could be turned around more effectively with polyphenols than with a scalpel. That revelation came from an overweight patient with blocked arteries who managed to clear up the blockages with a diet of exotic fruits and dietary supplements. Gundry was so impressed that he left surgery to, well, it is hard to say exactly what he has gone on to do. Frightening the public with dietary risks that lack evidence and selling supplements to reduce that imaginary risk would be one way to put it. Perhaps a pot had appeared at the end of a rainbow that was accessible without the drudgery of surgery.

To get our attention in the video, Gundry first tells us that he will start by revealing the three superfoods we should never eat. Never mind that "superfood" is a marketing term, not a scientific one. But before telling us what these three foods are, we have to listen to a litany of woes about how people today lack energy and struggle through the day and repeated urgings to make sure to watch the video to the end. After a few torturous minutes, we learn that the foods that we must avoid at all costs are soybeans, anything that contains lectins, and, of all things, goji berries. This comes after we were earlier told that polyphenols are the answer to all of life's problems. Well, soy contains a truckload of polyphenols, as do goji berries. There is no reason to avoid these foods. Lectins are a type of protein with various biological effects, some beneficial, some not. Digestion problems can crop up with overindulgence in lectin-containing foods, but since numerous grains and legumes contain lectins, avoiding them is a near impossibility and a non-necessity.

Lectins are extensively discussed in *The Plant Paradox*. Gundry is certainly right that these proteins are found in all sorts of foods. In both plants and animals, they perform a huge variety of functions, ranging from binding cells together to mediating immune reactions against invading microbes. When consumed in food, they are generally digested like any other protein and are not absorbed into the bloodstream, where some could indeed cause problems if present intact, as in the case of the injected ricin.

There are a few lectins that are difficult to digest and can create havoc by interfering with cellular activity in the digestive tract, causing some nasty symptoms that can include nausea, vomiting, and diarrhea. The classic example is phytohaemagglutinin, a lectin found in red kidney beans that can cause problems if it makes it to the small intestine intact. However, phytohaemagglutinin is readily destroyed by proper cooking.

In spite of a lack of evidence that lectins, aside from those in undercooked beans, are harmful, Gundry sells supplements that supposedly block the nonexistent nefarious actions of lectins. As with the plethora of other supplements he foists on visitors to his website, an asterisk leads to the disclaimer, in the tiniest of fonts, that "these statements have not been evaluated by the Food and Drug Administration; this product is not intended to diagnose, treat, cure, or prevent any disease." Then what, pray tell, are these products intended to do? Given that the pills cost $100 for a month's supply, a cynic might say that they are intended to make money for the marketer.

As far as the claims go, "help protect against infectious bacteria," "reduce your overall appetite," and "help relieve joint

pain" sure sound like claims of cures or prevention. Where are
the studies that show that Lectin Shield delivers the goods? We
are also told that this supplement can "help you get the body
you want, the energy you deserve, and the vitality you know is
inside — just waiting to break free." Give me a break.

Now back to the video and the sales pitch for polyphenols.
You would think that it was Dr. Gundry who discovered that
these compounds are found in food and that they may have
some physiological significance. In fact, polyphenols have been
extensively researched for over fifty years because of their antioxidant and anti-inflammatory effects, at least in the test tube.
Polyphenol intake may indeed be a reason people who eat lots
of fruits and vegetables are healthier than those who don't, or
the benefit may just be due to what they eat less of, like refined
grains or meat. Anyway, Gundry moans about how expensive
polyphenol-rich foods like berries are and how concentrated
extracts are the way to go.

Of course, none of the multitude of existing extracts meet
his exacting criteria. (Most likely because he is not profiting
from them.) But this savior has found a company capable of
producing a supplement that contains all the polyphenols an
energy-deprived person needs to restore vitality. He has seen
remarkable changes in his patients who have been doped up
on these supplements! Maybe so, but there is not one iota of
published evidence. The same goes for his metabolism-boosting
and probiotic products. Yes, there are anecdotes galore on Dr.
Gundry's website, but these don't amount to much. First, you
never know if they are authentic, and second, people who have
found no benefit do not tend to comment.

The interminable video then treats us to pictures of decrepit
seniors who are models for what will happen to us if we don't
invest in the Gundry supplements. And then comes the hard

sell with various discounts being offered if we order within the next few minutes (nonsense of course because the video is on the web to be viewed at any time) and great bargains to be had if we sign up for a lifetime supply. Gundry's zealotry for "Vital Reds" is distasteful. But a bowlful of berries does taste good, whether they are good for us or not.

The Plant Paradox begins with an introduction entitled "It's Not Your Fault," suggesting that whatever dietary efforts we may have been making to improve our health have likely failed because we have been fed the wrong information. But now, we are going to be set on the right track, the lectin-free track, by Dr. Gundry: "Suppose that in the next few pages I told you that everything you knew about your diet, your health, and your weight is wrong?" I would ask a question in return. "What is the likelihood that the thousands of researchers around the world, trained in biochemistry, nutrition, and epidemiology, who have spent careers trying to unravel the complex relationship between diet and health, have gotten it all wrong, and that a lone-wolf doctor with no relevant expertise has discovered the secret path to health?"

Let's just say that I will keep on eating my oats, whole grain bread, tomatoes, peppers, eggplant, beans, hummus, and fruits (in or out of season) with no reservation, since numerous studies support the benefits of such a diet. I may, however, look askance at someone pointing an umbrella at me.

WEIGHT LOSS SUPPLEMENTS

Universal agreement when it comes to nutritional issues is rare, but virtually everyone agrees that obesity is undesirable. There is no agreement, though, on how to address the problem. Infomercials, newspapers, and magazines tout a host of supplements guaranteed to "burn fat" or "rev up metabolism." Bookstore shelves sag under the weight of diet books, and researchers duke out the effects of intermittent fasting and the role of different dietary components in the pages of peer-reviewed journals. What does it all add up to for the consumer? Mostly confusion. But some of that can be dispelled just by asking a simple question: What is the evidence?

When it comes to dietary supplements for weight control, the answer is precious little. A typical newspaper ad sings the praises of a product that "curbs hunger," "burns calories instead of storing them," and "increases heat production by the body leading to weight loss." The heat is due to chili powder and the changes in metabolism are attributed to the inclusion of epigallocatechin gallate (EGCG), a compound found in green tea leaves. There is actually some evidence that EGCG at a dose of 200 to 300 milligrams a day can increase the oxidation of fats leading to an expenditure of about eighty calories, an insignificant amount.

Unfortunately, there is also evidence that EGCG can cause kidney and liver problems. It is all a question of the dose. The CBC's excellent program *Marketplace* undertook an investigation of green tea extracts after a couple of Canadian consumers developed liver damage, and the show's researchers turned up over sixty such cases documented in peer-reviewed literature. How much EGCG there is in supplements is generally a matter of some mystery, but toxic effects are possible even at 400 milligrams. The important point, however, is that there are no studies demonstrating that the particular chili powder and EGCG product advertised leads to weight loss. An anecdote that an unidentified woman went from size twelve to size six in six months and fit her wedding dress of thirty-seven years ago does not constitute evidence.

Evidence comes from peer-reviewed studies. But even in this

case, the evidence may not be of practical significance to people. A number of newspapers reported on a weight-control study with headlines like "Can Eating at the Same Time of Day Enable You to Burn off More Fat?" Yes, it can, if you are a mouse on a high-fat diet. Researchers hypothesized that the body clock could have an impact on metabolism and designed an experiment comparing mice with unlimited access to a high-fat diet with mice that had access to the same diet at set times during each day. The mice on the fixed eating schedule put on less weight even though their calorie consumption was about the same. An interesting finding, but mice aren't people and extrapolation of the findings to humans is unrealistic. While this study does not provide evidence that eating only at set meal times is conducive to weight loss, it can, however, raise that prospect.

David Zinczenko, who serves as the ABC television network's nutrition and wellness editor (despite having no background in science), has ideas along these lines. In his book *The 8-Hour Diet*, Zinczenko claims that the human body is designed for periods of eating and then periods of fasting and that consuming calories over an eight-hour period and then fasting for sixteen hours each day leads to weight loss. The hook for this regimen is that there are no restricted foods; anything can be eaten over the eight-hour period. But when Zinczenko's recommendations for foods that should be eaten are examined, it is clear that calorie restriction rather than the eight-hour time frame is the determining feature.

Michael Mosley, author of *The FastDiet*, also makes a case for intermittent fasting. Mosley graduated from medical school but became more interested in presenting science on television than practicing medicine. He has become an outstanding journalist, hosting a variety of programs including the BBC's excellent *Trust Me, I'm a Doctor*. In his book, he describes his 5:2 fasting diet, recounting his own experience, bolstered with theories drawn from the scientific literature. He suggests that consuming roughly 2,000 calories per day for five days a week and fasting at a level of 500 calories a day for two days not only leads to weight loss, but also to improved insulin sensitivity. Attention-grabbing, but in need of evidence from properly designed studies.

Unfortunately, where we do have plenty of evidence is in the failure of any kind of diet in the long run. The fact is that the majority of dieters end up regaining more weight than they lose. Still, they do not lose their appetite for the next scheme that comes along, be it the clay, tissue paper, or tapeworm diet. Don't ask.

ANTIOXIDANT HOOPLA

I stopped counting at about twenty but could have gone on and on. I was scanning product labels for the term "antioxidant" as I recently meandered through the Whole Foods supermarket in New York. There were juices from exotic berries, fruit concentrates, smoothies, an array of teas, dried plums, fresh blueberries, and powdered extracts of vegetables, grape seeds, and pine bark. And of course, innumerable capsules and pills loaded with vitamins, polyphenols, and minerals, all touting their potential to ward off disease and even boost life expectancy, thanks to their powerful antioxidant properties. What we have here, once again, is a modicum of science laced with a huge dose of marketing.

We can't live without oxygen. It is critical to respiration, the process by which cells use food components, mostly glucose, to produce energy. In an overly simplified fashion, electrons are removed from glucose, causing it to break down, eventually forming carbon dioxide that is exhaled. These electrons are transferred to oxygen via a complex series of reactions that allow it to combine with hydrogen to form water. The overall process leads to the release of energy and is basically analogous to burning a fuel, be it wood, coal, or gasoline. Fuel combines with oxygen to produce carbon dioxide and water with the simultaneous release of energy.

Our cells are actually little furnaces, and much like real furnaces, they produce side products, the notorious free radicals. Electrons extracted from food can go astray and add to oxygen molecules to produce superoxide, a free radical that can trigger a cascade of reactions forming other free radicals referred to as reactive oxygen species (ROS). These then can damage biomolecules such as proteins or DNA by ripping electrons from them. Such damage has been linked to cancer, heart disease, diabetes, arthritis, and cataracts, as well as the aging process. To complicate matters, free radicals are also produced on exposure to ultraviolet light, x-rays, ozone, or tobacco smoke. No surprise that by the 1980s, free radicals had been cast as molecular villains that had to be expunged. The question was how?

A clue surfaced from the numerous epidemiological studies documenting that a diet rich in fruits and vegetables reduces the incidence of the various diseases associated with free radical damage. Fruits and vegetables are composed of hundreds of compounds, many of which are capable of donating electrons to ROS, which neutralizes them. These are the "antioxidants," commonly portrayed as the heroes in the war waged against free radicals. Plants produce these compounds to protect themselves from oxidative damage caused by the oxygen they generate during photosynthesis. A seductive hypothesis now evolved. Squeezing more of these plant antioxidants into our body would allow them to act as sponges, mopping up those nasty free radicals.

Companies soon flooded the shelves with supplements containing vitamin E, vitamin C, beta-carotene, and a host of plant polyphenols. Cash registers jingled and people filled toilets with expensive urine since a good proportion of the antioxidants end up being excreted. That wasn't the major problem, though. The real problem was that properly controlled studies did not demonstrate a benefit from taking antioxidant supplements, and in some cases they even showed harm. Smokers actually increased their risk of lung cancer by taking beta-carotene! Maybe then, isolated antioxidants in pills aren't that useful, but what about those smoothies and other beverages that are touted as being especially high in antioxidants?

In a British trial, ten healthy volunteers were asked to avoid foods and beverages containing antioxidants for forty-eight hours before consuming a smoothie advertised as being rich in antioxidants. Blood samples were then periodically tested for antioxidant status during the next twenty-four hours. After an hour, there was an increase in antioxidant concentration but then levels quickly dropped below baseline, only getting back to normal after a day. So, even if antioxidants do neutralize free radicals, taking in more does not seem to be more effective since after an hour, the blood has a reduced antioxidant potential. This may be due to the body's tendency to deal with a sudden rise in blood components through homeostasis, the self-regulating process by which biological systems tend to maintain stability while adjusting to conditions that are optimal for survival.

Current scientific opinion is that the benefits of fruits and vegetables may be due to factors other than their antioxidant content and that free radicals may not be as villainous as previously believed. For example, subjects engaged in exercise training do not derive the benefit of increased insulin sensitivity when taking antioxidant supplements, possibly because the antioxidants neutralize free radicals that signal the body to respond to exercise. While high doses of free radicals are indeed dangerous, it seems that at low levels they induce a protective response to disease and aging. By all means, eat those fruits and vegetables, but cast a wary eye on the antioxidant hype.

MYSTERY OF THE GREEN BREAD

What is going on here? That's the text message I got from my daughter, accompanied by pictures of the "healthy" bread she had just baked. On slicing into the loaf, she was shocked to see that the inside had turned green! Was this safe to eat, she wondered? The mystery was solved as soon as I saw the ingredients that had been used. The batter included pumpkin, sunflower, chia and flax seeds, olive oil, honey, vinegar, coconut flour, almond flour, and baking soda. The culprits that gave rise to the green color were the sunflower seeds! I was quite confident of this because I was familiar with the problem faced by the sunflower oil industry in marketing the sunflower meal that is left over after the oil has been expressed from the seeds.

Sunflower oil sells well, mostly because of the high levels of oleic and linoleic acids, which being "unsaturated," are seen to be healthier than "saturated" fats. The large-scale production of sunflower oil means that an abundance of sunflower meal is available for sale. This is a nutritious commodity containing protein, fiber, minerals, and antioxidant polyphenols. The high protein content of the meal would make it useful as an ingredient in human food, but the stumbling block is that under conditions such as baking, the finished product develops a green color. This is completely harmless but causes consumer alarm with the result that sunflower meal is sold mostly as animal feed at a cost that is substantially lower than could be obtained were it to be sold as an ingredient for human food. As expected, much research has focused on the formation of the green color.

Curiously, the felon in the green color caper is also a nutritional hero. Chlorogenic acid is the main polyphenol found in the sunflower meal and has antioxidant properties that are generally seen as being desirable. Despite the "chloro" in the name, chlorogenic acid contains no chlorine. The name comes from the Greek for "light green" and a suffix meaning "give rise to." In the sunflower meal, chlorogenic acid reacts with itself to form a dimer, which in turn reacts with polyphenol oxidase, an enzyme released when the seeds are crushed to yield a compound that forms a green complex with proteins. Polyphenol oxidase is the same enzyme that is responsible for the brown discoloration of apples. The dimerization and subsequent oxidation of chlorogenic acid is enhanced under alkaline conditions, which now brings us back to the green bread.

Sunflower seeds were present in abundance, and baking soda provides the ideal alkaline conditions for formation of the green complex. Still, an experiment was needed to confirm the theory. I asked my daughter to immerse some of the sunflower seeds in a bicarbonate solution and set it aside for a while to see what happens. Sure enough, it wasn't long before the yellow seeds began to show a green tinge, turning completely green within a few hours. With heat, as during baking, the reaction would happen faster. The mystery of the green bread was adequately solved, and it was consumed without any anxiety. Even the grandchildren were treated to it.

It is not only with sunflower seeds that chlorogenic acid may lead to discoloration. It can cause potatoes to turn gray. But the chemistry is quite different from what happens in the sunflower case.

This time, the effect is due to a chemical reaction between

iron, a natural component of the potato, and chlorogenic acid. Iron in food can exist in one of two states, ferrous or ferric. In freshly cooked potatoes, a colorless chlorogenic acid-ferrous iron complex is formed, but upon exposure to oxygen in the air, it is oxidized to the colored ferric complex. One way of preventing this reaction is by using chemicals that bind iron and prevent it from reacting with chlorogenic acid. EDTA, gluconic acid, sodium acid pyrophosphate, and sodium bisulfite can do this and reduce the darkening. Commercial potato products such as frozen French fries, dehydrated mashed potatoes, and chips often are treated with sulfites, presenting a problem for people with a sulfite sensitivity. While this is not a true allergy, it can produce allergy-like symptoms including hives, swelling of the throat, cramps, and breathing difficulty. Sulfite sensitivity is rare in the general population, but roughly 10 to 20 percent of asthmatics react to sulfites.

Interestingly, the greening of baked goods made with sunflower seeds can also be prevented by the addition of sulfites, but that would not seem to be a wise use of additives. Why not just accept the greening and revel in understanding the neat chemistry involved? A great way to make holiday cookies without the need for artificial coloring!

BUTCHERING SCIENCE

"I think my butcher trying to put one over on me."

"What do you mean?" I queried the caller.

"I think he is packing fresh hamburger around old meat."

I began to sniff where this discussion was heading. "You mean when you dig into your packaged hamburger, you find that while the surface is red, the inside has turned brown?"

"Exactly!" came the reply. "Should I throw it out?"

It was time for a little chat about the chemistry of meat color. In a live animal, just like in humans, blood picks up oxygen as it passes through the lungs. Specifically, it is an iron ion embedded in a complex protein called hemoglobin that binds oxygen and delivers it to cells around the body. In muscle cells, hemoglobin transfers the oxygen to myoglobin, another complex protein that stores oxygen until it is needed. Every cell needs oxygen for respiration, the process by which glucose is "burned" to release energy.

Myoglobin is dark purple but converts to red oxymyoglobin when exposed to oxygen. The ratio of myoglobin to oxymyoglobin at any given time depends on the amount of oxygen available. When an animal is slaughtered, its blood is drained and no more oxygen is delivered to tissues. That is why the colour of freshly butchered meat is actually dark purple except on the surface, where it is contact with oxygen from the air. However, if the meat is now wrapped in a material that does not allow oxygen to pass through, the surface turns brown.

There is some interesting chemistry taking place here. The small amount of oxygen that still remains in the air surrounding the packaged meat, instead of binding to the iron in myoglobin, ends up stealing an electron from it. This reaction, which can be enhanced by the presence of bacteria, results in the formation of metmyoglobin, which is brown. This does not mean the meat is spoiled, although since it takes time for this reaction to occur, the appearance of brown color means the meat is not totally fresh. But this has no consequence when it comes to taste or safety.

What is the solution to keeping the surface from turning brown? Use packaging that allows oxygen to pass through, keeping the surface of the meat red. Polyethylene plastic wrap serves this purpose. But oxygen is a double-edged sword. It will keep the meat red, but it will also react with fat and cause rancidity and off-flavor. That's why polyethylene wrap is suitable for a few days but not longer. Although the surface stays red thanks to the formation of oxymyoglobin, very little oxygen diffuses into the meat. That little oxygen results in the formation of metmyoglobin, which is the reason that the inside turns brown. This is what was noted by my caller.

Neither should meat be frozen in its original polyethylene wrapper because this plastic allows moisture as well as oxygen to pass through. Loss of moisture results in "freezer burn." For freezing, meat should be wrapped in freezer paper that is coated with a special plastic layer such as polyethylvinylalcohol (EVOH) that is impervious to moisture.

As far as marketing goes, retailers would like to keep meat red for more than a couple of days in the refrigerator case. This is where modified atmosphere packaging (MAP) comes into play. In this case, plastics that are impervious to the passage of gases are used so that the headspace around the meat can be filled with a higher level of oxygen than is normally present in air. Such plastics usually are composed of different layers of polyethylene, EVOH, and sometimes PVC that prevents the oxygen from escaping, keeping myoglobin in its oxygenated form longer.

The extra oxygen also suppresses the growth of dangerous *Clostridium botulinum* bacteria, but it can also foster the growth of bacteria that prefer a high-oxygen environment such as pseudomonas. The growth of these bacteria, however, can be hindered by including carbon dioxide in the headspace. There is also a trick that can be used to keep meat looking red longer. When carbon monoxide reacts with myoglobin, it forms red carboxymyoglobin, so introducing this gas into the package will maintain the red colour. This is not allowed in Canada.

There is yet another method to keep meat looking red, one that is widely employed by producers of cured meat products such as hot dogs and cold cuts. When nitrite added to meat it is converted to nitric oxide, which then binds to iron in myoglobin to yield dark red nitrosomyoglobin. Heating converts this to nitrosohemochrome, the pink color typical of cured meats. Nitrite is a controversial additive given that it can react with naturally occurring amines in meat to yield nitrosamines, compounds that have been shown to be carcinogenic in animals. The doses used in these studies have been larger than what people may be exposed to, but epidemiological surveys have linked the consumption of cured meats to colon cancer. Whether this is due to nitrosamines or some other component in these foods is not clear, but in any case the high fat and salt content of cured meats is enough of a reason to limit their consumption.

Vacuum packaging is a way to keep meat fresh longer. It prevents microbial contamination from the outside and excludes oxygen to prevent fat oxidation. This also means that there will be no oxymyoglobin and the meat will be a dark colour. However, when removed from the packaging and cut, it will react with oxygen in the air and "bloom" to produce the red colour.

Packaged meat usually contains a small pad to absorb liquids released by the meat. This isn't only for cosmetic reasons. The liquid released by the meat makes for a good breeding ground for bacteria and can lead to contamination of the meat and even countertops. The polyethylene-lined absorptive packets can contain a variety of materials such as silica gel, plant fiber, or a superabsorbent polymer such as sodium polyacrylate.

After hearing about the colorful science behind meat packaging, my caller absolved her butcher of any questionable activity, but she was still "going to play it safe" and discard the meat. It seems you can lead people to science, but you can't make them digest it.

FOOD WRAPS

We feel a little guilt every time we bite into that hamburger, sub, pizza, or hot dog. They may comfort the taste buds, but the fat, salt, cholesterol, and sugar they contain are not welcomed by the rest of the body. There is yet another issue that arises. It involves the packaging in which these fast foods are delivered. The hamburger is wrapped in some sort of paper, fries come in a cardboard container, and that pizza may be delivered in a box. Since all these items are greasy and moist, ordinary paper won't do. It isn't very appetizing to pick up a hamburger wrapped in paper soaked through with grease. That's why various chemicals are used to provide moisture and grease resistance.

This, however, is not without controversy. Ideally, the packaging material should be both hydrophobic and lipophobic, meaning that it has the ability to repel both water and fat. That is quite challenging, but there is one class of chemicals, known as polyfluoroalkyl substances (PFASs), that is up to the task. Unfortunately, PFASs are also mired in controversy.

PFAS molecules are composed of a chain of carbon atoms to which fluorine atoms are attached. It is the fluorines on the periphery of the molecule that are responsible for repelling water and fat. But the presence of the carbon-fluorine bonds, which are very strong, also makes these molecules extremely resistant to degradation. The result is that they have been detected globally in water, soil, sediment, wildlife, and alarmingly, in human blood. Why the concern? Because some PFASs have been linked with thyroid disease, low birth weight, decreased sperm quality, higher cholesterol, as well as kidney and testicular cancer. It is important, however, to understand what "linked with" means.

While there is evidence that animals treated with polyfluorinated compounds exhibit the conditions mentioned, the doses are much higher than human exposure. Furthermore, exposure of animals to a single substance is a questionable model for extrapolating to effects in people who are exposed to literally thousands of compounds, both natural and synthetic, in their food on a daily basis. There are numerous interactions that can mitigate the effects seen with individual compounds. As far as humans go, high blood levels of PFASS, as can happen with occupational exposure, have been associated with disease. However, associations can never prove that there is a cause and effect relationship. For example, people may have high blood levels of these chemicals because they eat a lot of packaged fast foods, and it may be the diet that causes the problem.

Also, the specific molecular structure of the PFAS is important. Molecules that have a chain of eight carbons have been quite extensively studied and have been linked with health problems. Perhaps the most famous of these is perfluorooctanoic acid (PFOA), a compound that was widely used in the preparation of Teflon coatings before being phased out. It was not part of the final product, but during production, some PFOA was released into the environment. DuPont and its spin-off company Chemours recently settled thousands of lawsuits, to the tune of $670 million, with residents living around plants where DuPont used PFOA. The claim was that PFOA had poisoned the water supply and caused disease.

In North America, Europe, and Japan, these long-chain fluorinated compounds have been phased out, but because of their resistance to biodegradation, they are still widely present in the environment. Furthermore, they are still produced elsewhere, such as in China and Russia. In food packaging, they have been replaced by shorter chain fluorinated compounds that leave the body much more quickly and are less bioaccumulative. They still do show up in the environment, but their toxicity is thought to be less than their longer chain counterparts.

A 2017 study that received a great deal of publicity found that 46 percent of food contact paper, and 20 percent of food contact cardboard contains fluorinated compounds, although to what extent isn't clear. Nevertheless, this was enough to trigger headlines like "Not Only Can Fast Food Kill You . . . So Can the Wrapper." That is irresponsible journalism. The study did not investigate the extent to which any of these fluorinated compounds migrate into food, or whether they end up in the bodies of consumers. Detecting the presence of a substance does not automatically mean that it presents a risk. It is always a question of dose. Still, there is enough smoke around these fluorinated compounds to trigger a search for alternatives. There are some, like silicones and various hydrocarbons, but they are not as functional and also have raised some toxicity concerns. Obviously one way to cut down on exposure to chemicals used in packaging is to rely less on packaged fast foods. And that can pay other nutritional dividends as well.

IT'S IN THE CAN!

It may not be quite on par with the Manhattan Project or with the challenge of beating the Soviets to the moon, but the race to find a substitute for the lacquer used to line food cans is heating up. The canning industry is working on replacements for the epoxy resin currently being used because of concerns that bisphenol A (BPA), the chemical we have already encountered as an endocrine disruptor, may be leaching into the contents. BPA is combined with other components to form a polymer that keeps the metal from reacting with the food. Once the BPA has been incorporated into the polymer, it no longer has any hormonal effects, but there are always traces of unreacted BPA left over that can indeed leach out. Before exploring this issue, however, a bit of history is in order.

Napoleon, as many other generals before him, discovered that soldiers do not fight well on empty stomachs. And stomachs were often empty due to the difficulty of supplying food to massive traveling armies. So the emperor offered a prize of 12,000 francs, a healthy amount of money at the time, to anyone who could come up with a viable method of preserving food.

This challenge was taken up by Nicolas Appert, who, as the son of an innkeeper, had learned about brewing and pickling. He knew these fermentation methods could be halted by heat, and he began to wonder if food spoilage could also be stopped in this fashion. After all, it was clear that cooked food kept longer than fresh food, although eventually it too would spoil. Years of experimentation led Appert to make a critical discovery: if food was sealed in a glass jar and then heated, it would keep for a remarkably long time. Long enough to please Napoleon, at least, as he awarded the prize to Appert in 1809. The method clearly worked, although nobody at the time understood why. Bacteria were not identified as the cause of food spoilage until another famous Frenchman, Louis Pasteur, came along later in the century.

Appert's invention came to the attention of Peter Durand in England, who was troubled by the use of glass jars because they often broke. There had to be a better way! Why not a metal container? Iron was the first choice. But it would corrode, especially when exposed to acidic foods. A coating that would protect it from the air and contents had to be found. Tin, concluded Durand, would do the job! The metal had been known since antiquity and could easily be melted and applied to iron as a coating to make tin plate. And, most importantly, tin did not corrode. By 1818, the British company Donkin, Hall & Gamble was mass producing food in tin cans. When Admiral Parry sailed to the Arctic Circle in 1824, he and his crew subsisted on canned food. One can of roast veal apparently was not consumed, because it turned up in a museum 114 years later. Inquisitive scientists opened it and decided to check the effectiveness of the canning process. They were not quite brave enough to try the veal themselves, but the rats and cats that had the pleasure of partaking of the 114-year-old feast not only survived, but thrived!

Although tin did not corrode, small amounts did dissolve, resulting in tainted food. This also meant the possibility of forming microscopic holes through which bacteria could enter and undermine the canning process. Aluminum eventually turned out to be more suitable for cans but still presented the problem of the metal interacting with the food. Chemists now stepped into the picture and found that an epoxy resin made by reacting bisphenol A with epichlorohydrin was excellent for providing a barrier that was stable under the high heat and pressure of sterilization, did not crack if the can was dented, and stood up well to the varying acidity of different foods.

Epoxy resins performed admirably, but cracks, figuratively speaking, began to appear in the early 1990s. By then, analytical techniques had been developed to detect extremely small amounts of BPA, and more importantly, the hormonal effects of this chemical were being demonstrated by its effects on the multiplication of cultured breast cancer cells. In 1995, researchers at the University of Granada in Spain investigated a number of canned foods and found estrogenic activity in peas, artichokes, green beans, corn, and mushrooms, but not in asparagus, palm hearts, peppers, or tomatoes. The authors pointed out that while an estrogenic effect was observed, it was far less than that observed for estradiol, the body's naturally occurring estrogen.

The significance of the estrogenic effect of canned foods is difficult to estimate given that, on top of the estrogen produced by the body, as we have previously seen, we are exposed to a wide variety of natural estrogenic compounds found in foods that include milk, chickpeas, soybeans, vegetable oils, cabbage, flaxseeds, and oats. It should also be noted that the concentration of pure bisphenol A required to produce maximum proliferation of breast cancer cells in the laboratory is 1,000-fold greater than for estradiol.

Even though no risk from traces of BPA in canned foods has been demonstrated, there is clamor for invoking the precautionary principle, which aims to prevent harm even when the evidence is not fully in. For food companies, pleasing consumers is a high priority, whether consumers' demands are justified or not. So the race is on to find substitutes for epoxy resins. In some cases, for low-acid foods such as beans, plant extracts that harden into a resin have met with success. For other foods, companies are looking into various acrylics, polyesters, polyurethanes, and polyvinyl compounds. These do not match the performance of epoxy resin, nor is it clear that they have a better safety profile. Could we be trading in a perceived but unsubstantiated risk for a possible increased risk of food poisoning?

And one more thing: while you've been reading this little piece, hundreds of people have died from hunger, lack of clean water, poor sanitation, and a host of preventable diseases ranging from malaria to AIDS. By contrast, we have the luxury of worrying about traces of chemicals contaminating our ample food supply. A prescription for a dose of perspective is in order.

MOSCOW MULE

I have to admit that I had never heard of a Moscow mule until I came across this warning in the *Washington Post*: "Heads Up, Moscow Mule Lovers: That Copper Mug Could Be Poisoning You." It turns out that the reference is to a concoction of vodka, ginger beer, lime juice, and ice cubes that is traditionally served in a copper mug, probably to keep the beverage cold longer. Why Moscow mule? Nobody really seems to know, but the conjecture is that vodka is associated with Moscow and that the drink kicks like a mule.

In any case, the concern raised by the article is that acids in the drink can react with the mug and result in the leeching of copper compounds into the beverage that "could be poisoning you." Copper can indeed be toxic. But copper is also an essential nutrient, required by numerous enzymes in our body that are responsible for maintaining health. Of course, whether it harms or heals depends on the dose. We need an intake of roughly 1 to 2 milligrams a day, an amount easily met since copper compounds occur naturally in soil and water. Virtually everything we eat or drink contains trace amounts of this element. There is further exposure from water cruising through copper pipes and even from beer that is traditionally brewed in copper vessels to furnish yeast with the copper it requires for growth. Extensive research has shown that our daily intake of copper from all sources is about 2 milligrams per day, well below the 10 milligrams per day that can cause toxic effects.

The amount of leeching from copper containers depends on the acidity of the liquid with which the metal is in contact. Unfortunately, it seems that nobody has actually measured the copper content of a serving of Moscow mule, but it is known that pure water stored for sixteen hours in a copper vessel results in a concentration of 0.2 milligrams of copper per liter. Even if we assume that an acidic beverage like a Moscow mule leaches ten times more copper, that would only result in about 0.6 milligrams per 300 milliliters (10 ounce) serving. And of course the contact time would be nowhere near sixteen hours, meaning that we would be looking at a great deal less than that, nowhere near the 10 milligrams per day where toxic effects may occur. By this rather reasonable estimate, we can conclude that the amount of copper in a Moscow mule is hardly likely to poison you. In a bizarre twist, you can hop on the Internet and find a plethora of baseless advice to optimize health by drinking from copper mugs.

All this contemplation about leaching copper did something for me though. It triggered a childhood memory, one of a cherry-speckled pastry, in the angel food cake family, that my mother used to make. The recipe that was handed down to me describes beating egg whites into a firm foam before blending in sugar, flour, egg yolks, and fresh cherries. And the beating of the egg whites had to be done in a copper bowl! Why a copper bowl? Some interesting chemistry here.

A foam is basically a stabilized dispersion of gas bubbles in a liquid or solid. Beating egg whites causes the proteins they contain to unfold and join together to form a stretchable layer around the air bubbles that are introduced by whisking. This prevents the bubbles from coalescing. Recipes that recommend copper bowls for whipping the whites actually date back some 250 years, long before there was any understanding of what was going on at the molecular level. Julia Child was big on copper bowls.

It wasn't until the 1980s that food writer Harold McGee initiated experiments to determine what was going on. Using standard beating techniques, his experiments confirmed that egg whites whipped better in copper bowls than in glass bowls and the foam was more stable when sugar and flour were added as required by recipes. To prove that it was indeed dissolved copper that did the trick, further research showed that the addition of copper chloride to egg whites being whipped in a glass bowl achieved the same result as if they were beaten in a copper bowl. It wasn't a case of any metal. Iron did not do the trick.

Evidence suggests that copper ions form complexes with conalbumin, one of the main proteins in egg whites, basically making the protein film around the air bubbles more stable. The foam then becomes strong enough to withstand the assault of heavy sugar and flour particles being dropped on it. And it really does! I can confirm that because I actually did the experiment. I whipped the egg whites in a glass bowl and in a copper bowl and the foam in the copper bowl was decidedly firmer.

Copper bowls are not cheap, so if you do not own one, there is another trick that works. Just add a small amount of lemon juice or cream of tartar before whipping. This slight increase in acidity causes the long protein molecules to unravel more effectively, which in turn allows stronger interaction between them and that is just what is needed to form a network around air bubbles. What about the leaching of copper from the bowl? Based on the numbers from my Moscow mule estimates — totally insignificant.

For anyone still worried about being poisoned by drinking a Moscow mule from a copper mug, you can just use a mug that is lined inside with tin or steel, although aficionados claim that the taste is not the same. I would be willing to give that test a try without any concern for being poisoned by the copper. Actually, if there were to be any concern, it would be about alcohol, a chemical that is a proven carcinogen!

TILAPIA AND THE POOP CONNECTION

Fishermen tend to embellish the size of their catch, hence the expression "fish tale" for exaggerated stories like the ones making the rounds about tilapia, a fish that is increasingly showing up on dinner plates. Indeed, it is now the most widely consumed fish after salmon and tuna. Typically, headlines scream about tilapia being "Poop Fish," "Worse Than Bacon," "No Better for You Than a Donut," and that it is "Like Eating a Rat!!" Relax. Tilapia will not poison you. You are better off eating it than bacon or donuts. As far as rats go, there are no studies on their nutritional value since few humans make a habit of dining on the rodents. But I suspect tilapia tastes better.

The increasing popularity of tilapia is due its mild taste and the relative ease with which the fish can be raised on fish farms, leading to a lower cost. Although there are tilapia farms in North America, most of the tilapia consumed are imported from Asia, with China being the main producer. The poop connection arises from some unscrupulous operations that have been identified there. Tilapia in the wild feed on algae, but on farms they are reared on corn or soybean meal. However, when no other feed is provided, they will eat poop. There have been instances where fish farms in Asia were found to be feeding poultry, sheep, or hog manure to tilapia. While this does not mean that eating these fish is tantamount to eating poop, the practice does increase the risk of bacterial contamination and the need to treat the fish with antibiotics. It isn't clear just how widespread this practice is in Asia, but it doesn't occur in North America where the quality of the water in which tilapia are raised is also carefully monitored.

The nonsensical "worse than bacon" and "worse than donuts" stories can be traced to a publication in the *Journal of the*

American Dietetic Association back in 2008 in which researchers from Wake Forest University reported on determining the ratio of omega-6 to omega-3 fatty acids in a variety of fish. This is of interest because omega-3 fats can reduce the risk of irregular heart rhythms, ease inflammation, and inhibit the formation of blood clots while the omega-6 versions have been linked, albeit somewhat controversially, with increased inflammation, a contributing factor to chronic conditions such as heart disease and diabetes. In theory then, the lower the ratio of omega-6 fats to omega-3s, the less the likely the food is to trigger inflammation. Based on this view, the Wake Forest paper somewhat unfortunately noted that "all other nutritional content aside, the inflammatory potential of hamburger or pork bacon is lower than the average serving of farmed tilapia." As one might expect, this observation spawned a number of scary media accounts.

Of course, it is totally unrealistic to "put all other nutritional content aside." Hamburger and bacon contain far more saturated fat and cholesterol than tilapia and are burdened with other issues such as nitrite content and the formation of various carcinogens on cooking. Furthermore, the role that omega-6 fats play in health is complicated. The main omega-6 fat in the diet is linoleic acid, which in the body is converted into arachidonic acid that then is converted into a variety of compounds, some of which promote inflammation while others have an anti-inflammatory effect. Also to be considered is the ability of omega-6 fats to reduce LDL, the so-called "bad cholesterol" and boost HDL, the "good cholesterol." Indeed, some studies suggest that the decline of heart disease in North America since 1960 can be attributed to replacing saturated fats with omega-6 fats. Basically then, the degree to which the omega-6/omega-3 ratio is to be regarded as a significant risk factor is unclear.

As far as donuts go, their sugar and saturated fat content trumps any worry about unsaturated fat ratios. It is also noteworthy that many "healthy foods," such as nuts and grains, have a considerably greater omega-6/omega-3 ratio than tilapia. There is far more to determining the "healthiness" of our diet than this ratio. While it may be meaningful in the context of the overall diet, it doesn't have much meaning when looking at individual foods.

It is true that wild tilapia eating algae rich in omega-3 fats have a more favorable ratio than farmed fish that are fed a diet of corn and soy in which omega-6s predominate. Like people, fish are what they eat. It is also correct to say that other fish, such as salmon and tuna, have far more of the beneficial omega-3s than tilapia. On the other hand, since tilapia do not eat smaller fish, they have a lower mercury content than most other fish since mercury gets concentrated up the food chain.

Given that male tilapia grow larger than female, they are more profitable to produce. Interestingly, tilapia are actually born genderless and can be made to develop into males with the addition of methyltestosterone to their diet for a short time after birth. By the time the fish are marketed at the age of six months, there is no residue of this hormone in the flesh. There is some concern about releasing wastewater from facilities where hormones have been used, but gravel and sand beds can filter out any methyltestosterone. In any case, in North America such hormone treatment is rarely used.

The bottom line is that farmed tilapia can certainly fit into a healthy diet and provide a good alternative to meat. If a choice is available, North American tilapia is a better bet than fish that are imported from Asia. Given that in about twenty-five years, we will be looking at a world population of nine billion, fish farming is going to take on greater and greater importance. It is unfortunate that some people will shy away from eating tilapia because of the clutter of fish tales about its risks to health.

JELLYFISH PROTEIN AND BRAIN FUNCTION

Wouldn't we all like to have healthy brain function, a sharper mind, and clearer thinking? Of course we would. And a dietary supplement called Prevagen promises to deliver the goods, at least according to the ads that are featured in magazines and on CNN ad nauseum. The supposed active ingredient is a protein called apoaequorin that according to the marketer "was originally plucked from a variety of jellyfish." That bit of info is meant to conjure up an image of safety, catering to the notoriously false belief that natural substances are somehow inherently safer than synthetic ones. Never mind that some jellyfish actually produce a venom that can be lethal to humans. Of course the origin of apoaequorin is irrelevant when it comes to safety or efficacy. What matters is what the evidence demonstrates.

Why should there be any connection between a jellyfish protein and brain function? The glow produced by some jellyfish is produced when apoaequorin binds to calcium, a finding that is of interest to researchers because the human brain also contains calcium-binding proteins that play a crucial role in brain function. Nerve cells need calcium for proper functioning, but it has to be just the right amount of calcium, not too little and not too much. Calcium-binding proteins protect against excess calcium, but unfortunately with age the levels of these proteins decrease, and the resulting high levels of unbound calcium can damage nerve cells, impairing thought and memory. The jellyfish protein has an amino acid sequence that is very similar to that of the body's calcium-binding proteins, and indeed, when cells in the lab are treated with this protein, they are more resistant to induced damage. This has spawned the idea that apoaequorin can protect nerve cells from some of the consequences of aging. Maybe so, in cell cultures in the lab.

However, if this protein is ingested in a pill form, for any hope of efficacy it would have to survive digestion, be absorbed into the bloodstream, and then cross the blood–brain barrier. But proteins are readily broken down into amino acids and peptides during digestion, so the chance of an intact protein ending up in brain cells is remote, to say the least. Forget the scientific implausibility though; the relevant question is whether Prevagen works.

Quincy Bioscience, the company that markets Prevagen, hypes an in-house placebo-controlled trial that shows memory improvement after ninety days. Actually, while there were some positive results in a few specific tests, overall the placebo group performed as well as the experimental group. Aside from the company's own questionable trial, there are no peer-reviewed publications attesting to the safety and efficacy of Prevagen. Nevertheless, Quincy Bioscience claims that its product has been clinically shown to improve memory and provide cognitive benefits, although the small print in the ad does state that "these statements have not been evaluated by the Food and Drug Administration. This product is not intended to diagnose, treat, cure, or prevent any disease."

There is another kicker. The company's "evidence of safety" basically admits that the product cannot work. The unpublished, hyped study concludes that "the current study assesses the allergenic potential of the purified protein using bioinformatic analysis and simulated gastric digestion. The results from the bioinformatics searches with the apoaequorin sequence show the protein is not a known allergen and not likely to cross-react with known allergens. Apoaequorin is easily digested by pepsin, a characteristic commonly exhibited by many non-allergenic dietary proteins. From these data, there is no added concern of safety due to unusual stability of the protein by ingestion." In other words, the protein is broken down in the digestive tract and does not enter the brain.

To complicate the situation further, the Food and Drug Administration in the U.S. has charged that Quincy Bioscience is marketing Prevagen as a dietary supplement, a category for which it does not qualify since apoaequorin is synthetically produced. Based on the claims made for the product, it should be classified as a drug, but it has never been approved as such. That would require a degree of evidence that the company cannot provide. The FDA also claims that the company has not disclosed over a thousand reported adverse reactions to Prevagen, including seizures, strokes, and worsening symptoms of multiple sclerosis as well as chest pains, tremors, fainting, and curiously, memory impairment and confusion.

Little wonder that consumers are not enamored of Prevagen. Some have even organized a class action lawsuit against the company claiming that "Prevagen is a singular purpose product: its only purported benefit is to enhance brain function and memory — which it does not do."

One supposes the company's executives take their own product. Their lack of clarity about evidence-based science hardly supports the purported benefits of Prevagen.

HAWKING BRAIN SUPPLEMENTS

Dr. Stephen Hawking was a brilliant theoretical physicist. But even brilliant scientists can be wrong, as Hawking was about his contention that the Higgs boson, an elementary particle, the existence of which had been theoretically predicted, would never be found. To his credit, after years of public debate with Peter Higgs, he quickly acknowledged that he had been wrong when the particle was discovered in 2012. He went on to say that Higgs should receive the Nobel Prize, which he did in 2013.

Hawking had strong views on climate change and in the years before his passing expressed outrage at U.S. environmental policies. He was also a great supporter of stem cell research, space exploration, and potential human colonization of other planets. Rare for a scientist, Professor Hawking achieved a celebrity status that ensured his comments on issues ranging from universal health care and Brexit to Scottish independence and Middle East politics received media attention. So it comes as no surprise that a quote attributed to Hawking about the dietary supplement Inteligen has also received publicity. "This Pill Will Change Humanity" is certainly an alluring proclamation, but the fact is that Hawking never said it!

Unscrupulous marketers have seized upon Hawking's fame to push Inteligen. The caption on what looks like a screen image captured from CNN reads, "We can now access 100 percent of the brain," and features a picture of Hawking. The story goes on to say that during an interview with Anderson Cooper (although strangely the picture shows Wolf Blitzer), the famous physicist said that his brain is sharper than ever, more clear and focused, largely because he has been taking Inteligen. He is quoted as having claimed that "the brain is like a muscle, you got to work it out and use supplements just like body builders use, but for your brain, and that's exactly what I've been doing to enhance my mental capabilities." Needless to say, Hawking never said any such thing and was never interviewed by Anderson Cooper (or Wolf Blitzer). Furthermore, although he was in possession of a dazzling brain, Hawking had no expertise when it came to drugs that claim to enhance brain function, so his comments wouldn't carry much weight even if he had made them, which he didn't.

The fake news story directs to websites where Inteligen can be purchased. Some of these appear to be fake as well, designed to snare people into providing their credit card information. Others offer information about the product, claiming it reduces mental fatigue, dispels brain fog, and improves memory. There is no official website for the product, no patent for Inteligen has ever been filed, and no clinical trials have been registered. Curiously, the various websites that promote the supplement can't agree on what the ingredients are, although most mention *Bacopa monnieri*, vinpocetine, *Ginkgo biloba*, and acetylcarnitine. None provide any info about how much of these are present, which of course makes it impossible to evaluate just what the effectiveness of this brain supplement is, if it actually does exist.

Bacopa monnieri is a plant that has historically been linked to cognitive function in traditional medicine. Like any plant, it contains numerous compounds, including "bacosides," compounds shown to protect nerve cells from oxidative damage in animal models by regulating antioxidant enzymes such as superoxide dismutase and catalase. At least one recent study in human subjects indicated an improvement in some aspects of memory, but there were also reports of abdominal cramps and nausea. Nevertheless, that study was enough for Dr. Oz to jump on the bandwagon and recommend *Bacopa* to enhance memory, improve focus, and to make one smarter. Whether he takes it himself is not known.

The *Ginkgo biloba* tree also has a long-standing connection with cognitive improvement that traces back to ancient China when nobles treated themselves with the nuts of the tree, hoping to prevent senility. Today, various dietary supplements sourced from the tree are promoted with claims of slowing cognitive decline and enhancing memory. Numerous studies have failed to provide evidence that *Ginkgo* has any significant effect on brain function or that it is helpful for any health condition.

Acetylcarnitine is a naturally occurring compound in the human body derived from amino acids in the diet. It has a purported ability to increase production of the neurotransmitter acetylcholine, a compound that does play a role in brain function. Some studies using supplements of acetyl-L-carnitine have shown some mild improvement in cognitive function but only at doses of 2 grams a day, far more than can conceivably be present in Inteligen.

Now we come to vinpocetine, "a chemical from the periwinkle plant" that is widely advertised to increase blood circulation to the brain and to act as "a cognitive protectant." Animal studies have indeed shown that vinpocetine can reduce the loss of nerve cells due to decreased blood flow and a few placebo-controlled studies in the elderly have shown some improvement in concentration and memory. But here is the glitch. Vinpocetine does not actually occur in the periwinkle plant! However, a compound called vincamine that can be converted in the laboratory to vinpocetine, does. Saying that vinpocetine "comes from periwinkle" is an attempt to capitalize on the false but popular notion that natural substances are safer than synthetics. Vinpocetine is not a vitamin, mineral, amino acid, or a botanical substance and therefore should not fall into the category of a dietary supplement. It is a synthetic compound for which health claims are made and should therefore be considered a drug.

When it comes to Inteligen, the intelligent questions to ask are: Do we know what the ingredients really are? Do we know how much of each ingredient is present? Have any clinical trials of the product been carried out? Is it being sold by a reputable company? And did Stephen Hawking really endorse this supplement? The answers to all is "no." I'm sure Dr. Hawking would have agreed.

ADVANCED GLYCATION END PRODUCTS

You scrutinize food labels for cholesterol, sodium, sugar, and trans fats, the usual suspects for playing mischief with health. Let me now add a little more drama to your life by introducing yet another potential culprit, one you will not find listed on any label. Advanced glycation end products, or AGEs, are formed in cooked or processed foods when amino acids or fats react with sugars. This wide array of compounds can enter the bloodstream upon ingestion and predispose individuals to oxidative stress and inflammation, factors that are believed to play a role in causing many chronic conditions, including type 2 diabetes, cardiovascular disease, arthritis, cataracts, kidney problems, cancer, and perhaps even Alzheimer's and Parkinson's disease. To complicate matters, AGEs can also form in our body and accumulate over time to play a role in aging.

Before getting down to details about AGEs, a little clarification is in order as to why oxidative stress and inflammation should be a concern. Think of the body as a laboratory in which myriad chemical reactions — ranging from enzymes targeting potential toxins to oxygen combining with glucose to produce energy — are going on all the time. Some of these produce debris in the form of free radicals, highly reactive oxygen species that can damage proteins and DNA. Although our bodies have evolved various ways of neutralizing the reactive oxygen species, oxidative stress can occur when there is an imbalance between the production of free radicals and the ability to detoxify them or to repair the resulting damage. Such stress can happen when the body launches a chemical attack to eliminate AGEs, resulting in "friendly fire" in the form of free radicals.

Inflammation, the body's way of initiating a healing response, is actually related to oxidative stress. If a cut or scrape swells, and gets red, hot, and painful, it is because immune cells rush to the area to deal with microbes or toxins that now have an entry pathway into the bloodstream, often using free radicals as weapons. Once the intruders have been dealt with, the inflammation subsides, but inflammation can also occur inside the body in response to potentially harmful substances. If these are continually present, as can be the case with AGEs, constant low-level inflammation ensues. As white blood cells gather to try to eliminate the intruder, they can end up damaging healthy tissue as well.

Reducing oxidative stress and chronic low-grade inflammation is obviously desirable, and since AGEs have been implicated in these conditions, attempts to reduce exposure seem appropriate. Underlining this is the finding that AGEs may cause damage by another route as well. They can cross-link proteins, altering their properties and function.

Although the proposal of a link between advanced glycation end products and disease is recent, the reaction by which AGEs form has been studied for over a hundred years. It was back in the early twentieth century that French physician Louis Maillard, who had a particular penchant for chemistry, became interested in why foods like bread or meat turned brown on baking or cooking. He concluded that heat caused amino acids and fats in these foods to react with sugars, a process referred to as glycation. The Maillard reaction, as it came to be known, eventually turned out to be extremely complex as the initially formed glycation products were found to undergo a variety of further reactions resulting in the advanced glycation end products. The key factor in their formation is dry heat.

Evidence for the possible damage caused by AGEs comes both from animal and human studies. Mice have elevated circulating and tissue AGEs when fed a diet high in these substances and also exhibit a greater risk of atherosclerosis and kidney disease. When placed on an AGE-restricted diet, they show improved insulin sensitivity and live longer. Human evidence parallels these findings. Researchers at Mount Sinai Medical Center in New York randomized obese subjects into two groups, with one consuming a low-AGE diet and the other a standard American diet with high AGE content. After a year, the first group had lower blood levels of AGEs, a reduction in insulin resistance, and a decrease in markers for inflammation and oxidative stress.

Now that I have your attention about a substance that you probably didn't even know exists, it is time for some good news. Regimens that reduce dietary AGEs and also curb their production in the body are the ones that are recommended anyway for the maintenance of health. Since sugar is a key ingredient in the Maillard reaction, reducing it in the diet will lessen the chance of AGEs forming in the body. It is no surprise that diabetics who have elevated sugar levels are more prone to AGE-associated diseases.

Most significantly, modifying cooking methods can effectively reduce exposure to AGEs. Foods rich in both protein and fat, mostly animal products, will contain lots of AGEs when broiled, grilled, fried, or roasted, with many fast and processed foods falling into this category. Low-temperature moist cooking, such as stewing or microwaving, results in low AGE content. Diets rich in fruits, vegetables, whole grains, legumes, fish, and low-fat dairy, the Mediterranean diet being an example, are typically low in AGEs.

My preferred meat is chicken, and I alter between grilling, frying as schnitzel, or stewing as chicken paprikash. In light of the AGE revelations, I now gravitate more towards the paprikash. An added benefit here is that the tomatoes, green peppers, onions, and paprika also provide antioxidants that have been shown to mitigate the production of AGES. When I made my last batch, just for fun I put on a chef's hat that prompted my daughter to quip that I looked like a giant Q-tip. I'll put that comment to use. Q: How do I reduce AGES? Tip: If you stew you won't accrue.

SNIFFING FOR FENUGREEK EFFECTS

If you have eaten curry, you have probably tasted fenugreek. The seeds of this plant as well as its fresh leaves are commonly used as ingredients in curries. They are added for taste, but they also impart a smell that is due to sotalone, a compound that at low concentrations has a distinct maple syrup–like odor. Since sotalone passes through the body unchanged, it can impart a scent both to the urine and sweat.

It is not only curry eaters who smell of maple syrup. It can be an issue for lactating mothers who take fenugreek supplements to increase milk production. While there is much anecdotal evidence that this works, the few studies that have been carried out have shown mixed results. There is always a question of just how much to take, which is tough to answer because herbal supplements are difficult to standardize and often there is a mismatch between what is indicated on the label and what is actually in the product. Some women take blessed thistle along with fenugreek because this herb also has a reputation as a lactating agent. In this case, there is insufficient evidence for efficacy or about the safety of taking this herb during pregnancy or while breastfeeding. Blessed thistle is not the same as milk thistle, which in spite of its name has nothing to do with encouraging milk production.

Herbal remedies are drugs and like any drug can have side effects. As a food, fenugreek rarely causes problems, but as a supplement it can result in loose stools and intestinal discomfort. Allergy to fenugreek is possible, especially in people who have allergies to peanuts and chickpeas, which are in the same botanical family. Since fenugreek can lower blood glucose, it can in some cases cause hypoglycemia. This is of special concern in diabetics because fenugreek may enhance the effect of anti-diabetic drugs. On the other hand, fenugreek's blood glucose lowering effect may be of some benefit to diabetics, but again there is the problem of knowing how much to take because of lack of standardization.

Stimulated by anecdotes of diabetes control with fenugreek in India, researchers have begun to investigate the potential therapeutic use of the plant. Animal trials have clearly shown that fenugreek can lower blood sugar levels and the few human trials that have been carried out lend support to the proposed hypoglycemic (blood sugar–lowering) effect of the seeds of the plant. Both the powdered seeds and water-alcohol extracts of the seeds have been investigated. A placebo-controlled trial in which type 1 diabetics were given 50 grams of defatted fenugreek seed powder twice daily did demonstrate a blood glucose lowering effect, but the trial involved only ten patients. A larger study of sixty patients with type 2 diabetes used 25 grams of powdered seed twice daily and showed a decline in fasting blood glucose from 8.4 millimoles per liter to 6.2 millimoles per liter. The average A1C, a measure of blood glucose control over a period, declined from 9.6 percent to 8.4 percent. A small trial using 1 gram of a seed extract per day also showed improvement in glucose control. But fenugreek also contains a small amount of coumarin that can interfere with anticoagulant medications. All in all, there is some evidence that powdered fenugreek seeds, in the ballpark of 25 to 100 grams a day, may help with the control of blood glucose, but the evidence is not compelling.

Why fenugreek should have any effect on blood glucose isn't clear. As any plant product, fenugreek seeds contain dozens of compounds including 4-hydroxyisoleucine, which may stimulate insulin secretion. There are also steroids such as yamogenin and diosgenin, which may explain its reputation to induce labor and help with menstrual complaints. Indeed, in the nineteenth century, Lydia Pinkham used fenugreek seeds in her magical cure-all for "female problems." Men were not forgotten either. There were unfounded claims about fenugreek increasing libido and being of help in erectile dysfunction. And long before, the ancient Egyptians obviously considered fenugreek to be important since seeds were found in King Tutankhamen's tomb, supposedly to help sustain him in the other world.

Since fenugreek can cause uterine contractions, supplements should not be taken during pregnancy. When taken for lactation, the advice that is often offered is to slowly increase the dosage until the sweat or urine begins to smell like maple syrup. Breastfed babies may also smell of maple syrup if the mom has been ingesting fenugreek, and that can lead to false diagnosis of "maple syrup urine disease." This is a serious genetic disorder characterized by a deficiency in enzymes that metabolize the common amino acids valine, leucine, and isoleucine. A buildup of these amino acids and their breakdown products can lead to severe neurological damage and eventually to death. One of these breakdown products is sotalone, so its odor can indicate maple syrup odor disease. Today, should the condition be suspected based on a baby's failure to thrive, testing of the blood amino acids can detect the condition even before any scent appears. Serious consequences can then be avoided by adhering to a diet that is based on a special formula free of the problematic amino acids.

Interestingly, sotalone can be isolated from fenugreek seeds and is actually used as one of the flavor components in artificial maple syrup. Facilities that process the seeds often smell strongly of maple syrup, and the scent can be carried quite some ways by the wind. Back in 2005, Manhattanites began to complain of a strong maple syrup odor and rumors circulated about it being some sort of chemical warfare. It took a while, but eventually the smell was traced to a company in New Jersey that was processing fenugreek seeds. That rumor even made it onto an episode of *30 Rock*, the popular sitcom. And eventually into this book.

QUESTIONING PHOSPHATES

Fried bacon. A grilled cheese sandwich. Self-basting turkey. A muffin. Ice cream. A can of tuna. A cola drink. What do they have in common besides frequently cropping up in the Western diet? They all contain some form of phosphate additives. And that is turning out to be an issue. Too much phosphate in the diet may have health consequences, particularly as a risk factor for heart disease.

This possibility may be surprising because phosphates are also essential for health. They are a component of DNA, the "blueprint of life," as well as of adenosine triphosphate, ATP, the key molecule in the cell's production of energy. Phosphates are components of the phospholipids that make up cell membranes, and calcium phosphate is the main building block of bone. However, it's simple to meet the body's need for phosphates because they occur naturally in plants and animals. Plants get their phosphate from the soil, and animals get them from plants. Soil phosphate has to be replenished, making phosphate fertilizers produced from phosphate-containing rocks an essential feature of modern agriculture. But they are not without problems. When phosphate washes into lakes and rivers, it increases plant growth, and when these plants die, they decompose and use up the dissolved oxygen in the water, robbing fish of the oxygen they require for life.

Phosphates in plants are integral components of molecules such as phospholipids and DNA and are liberated during digestion to be absorbed and used to build the body's requisite phosphate-containing compounds. We have evolved to absorb only the amount of phosphate needed; the rest is eliminated in the urine and feces. Indeed, the first problem with phosphates cropped up in patients with kidney problems who suffered from a high rate of cardiovascular disease that was traced to high serum levels of phosphate, a consequence of their kidneys being unable to filter excess phosphate. Excess phosphate combines with calcium in the bloodstream to form deposits of calcium phosphate in arteries, leading to a "hardening" of these vessels. The prevailing opinion has been that this is only a problem for kidney patients, who therefore have to maintain a low-phosphate diet because their kidneys cannot clear phosphates efficiently.

However, recent studies have shown that excess phosphate may also be a problem for people with normal kidney function. One study even found hardening of arteries in young men with high serum levels of phosphate. This seems to be the result of the high phosphate levels reprogramming smooth muscle cells in blood vessels to become bone-forming cells. The puzzling question is how people with normal kidney function end up with high levels of phosphate. The answer appears to be the increased use of phosphate additives in food.

Since the 1990s, our intake of phosphates in the form of food additives has doubled, from 500 milligrams a day to 1,000 milligrams. And these phosphates are much more readily absorbed than naturally occurring phosphates and are more likely to lead to elevated phosphate levels in the blood.

Phosphate additives are everywhere because they are very useful. Monocalcium phosphate, for example, is used in baking powder to generate carbon dioxide from sodium bicarbonate, meaning that many baked goods contain phosphates. Tricalcium phosphate is added to salt to keep it free flowing, and disodium phosphate reduces cooking time of cereals. It is also used as an emulsifier in processed cheese, preventing butterfat from separating. Tetrasodium dihydrogen phosphate acts together with calcium to link milk proteins together and is the secret behind thick milkshakes. Sodium acid pyrophosphate keeps potatoes from turning gray when boiled, and phosphoric acid is added to cola beverages as a flavoring and to prevent the caramel color from turning too dark.

An important property of phosphates is that they can sequester or bind metal ions like iron, copper, and magnesium that are needed by bacteria for growth and also catalyze oxidation reactions. By binding these ions, phosphates act as preservatives. Canned seafood can form crystals of magnesium ammonium phosphate, known as struvite, that can look like tiny slivers of glass. Their formation can be prevented by phosphates that tie up the magnesium.

Perhaps the property that food producers favor the most is the ability of phosphates to bind water. Injecting ham, hot dogs, cold cuts, or turkey with phosphates leads to a juicier product, but it also allows producers to sell water as if it were meat. These foods will declare phosphates on their labels, but in some cases, they may be used without the consumer's knowledge. "Purge" is a liquid released from raw meat as it ages. Consumers shy away from packages with high purge, resulting in reduced sales and increased waste. Treating the meat with phosphates reduces purge, resulting in increased sales.

Phosphates have also been accused of contributing to attention deficit hyperactivity disorder (ADHD) in children. This allegation is on much weaker footing than the claim that phosphates play a role in cardiovascular disease. There's no real evidence here other than parents claiming that their hyperactive children improve on a low-phosphate diet. This may be due to the fact that a low-phosphate diet means eliminating processed foods, or it may be something else in these foods that is triggering the hyperactivity. In any case, there is no doubt that commercially processed food is much higher in phosphates than fresh food, so a drive to reduce phosphates will lead to a cutback on processed foods, which is desirable for a variety of reasons.

If you would like a visual reminder of phosphate additives, next time you watch your bacon fry, note the shrinkage as the water that was bound by phosphates is released. The phosphates stay behind, along with the salt and nitrites. Bacon is no health food. But you could have guessed that since it tastes too good.

A TALE OF TWO CANTALOUPES

This is a tale of two cantaloupes — one that killed and one that cured. Herb Stevens was a spry eighty-six-year-old who suddenly developed tremors and chills and became so weak that he was unable to get up from the toilet. And so began a downward spiral of complications that would eventually lead to his demise. Tests revealed that Mr. Stevens had been infected with *Listeria monocytogenes*, a soil bacterium commonly found in animal feces. Two weeks earlier, the retired hydrologist had eaten half a cantaloupe purchased at a local Colorado supermarket, a purchase that would turn out to have lethal consequences.

Unfortunately, Mr. Stevens was not the only victim; before the 2011 *Listeria* epidemic subsided, 147 people would be hospitalized and thirty-three would lose their lives. All had eaten cantaloupes that were eventually traced to a Colorado farm owned by brothers Eric and Ryan Jensen, who were charged with introducing adulterated food into interstate commerce. The charge did not imply that they knew, or should have known, about the contamination, but as owners of the farm, they were accountable for maintaining sanitary conditions. Prosecutors urged a heavy-handed approach to send a strong message to the food industry about its responsibility to reduce food-borne illness. It is indeed a critical responsibility, given that bacteria and viruses lurk everywhere in our food supply, just waiting for a chance to wreak havoc on our health. The judge in the case agreed with the prosecutors, sentencing the Jensens to five years of probation and six months of home detention. Each also was ordered to pay $150,000 in restitution and perform 100 hours of community service.

It is difficult to estimate the extent of illness caused by microbes because the vast majority of cases resolve after a brief tussle with cramps, nausea, and diarrhea, and never get reported. The so-called twenty-four-hour flu is a misnomer. Influenza is not a one-day phenomenon, but symptoms associated with food poisoning can sometimes pass in twenty-four hours. If you're lucky. If you're unlucky, contaminated food can kill. The Centers for Disease Control and Prevention in Atlanta estimates that in the U.S. there are about 50 million food-related illnesses a year, with 130,000 hospitalizations and 3,000 deaths. Most of the victims are children, the elderly, and people whose immune systems are compromised. Pregnant women are especially at risk, but healthy people generally are not seriously affected.

Virtually any food can be contaminated by bacteria, but cantaloupes are particularly prone because of their continued contact with the soil during growth. Furthermore, their rough skin can trap and hold bacteria, some of which can even penetrate to the inside of the melon. Just slicing a melon can transfer bacteria from the outside to the inside, which is why washing the fruit before cutting is wise. Growers are expected to minimize risk by maximizing precautions, but there are many points during processing that allow for the possibility of contamination. Investigation showed that Jensen Farms was lax in the maintenance of proper sanitary conditions.

Although the exact source of the bacteria was never pinpointed, numerous samples taken around the farm revealed the presence of *Listeria*. Contact with a hard-to-clean potato-washing machine was a possibility, as was cattle manure found on cantaloupe transport truck tires. The packing house had pools of water on the floor, and the melons were not properly cooled after coming off the fields.

Listeria is not the only bacterium that can contaminate cantaloupes. In 2012, a *Salmonella* outbreak that made 261 people sick and caused three deaths was traced to Chamberlain Farms in Indiana. The victims were located in twenty-four states, a sobering reminder of how our current food distribution system can cause widespread problems. A cattle pasture next to the growing field may have been the origin of the bacteria, but once again the major problem was a bevy of inadequate practices in the packing house that ranged from lack of monitoring of washwater disinfectant levels to droppings from birds roosting in the rafters above food contact surfaces.

These outbreaks, although tragic, highlight the need for vigilance in food production. At home, wash produce with running water even if it is going to be peeled. Fruits or vegetables with uneven surfaces, such as cantaloupes, can be scrubbed with a produce brush to remove microbes that are otherwise difficult to dislodge. Care should be taken not to spray water during washing because bacteria and viruses can live on surfaces for a long time. For extra safety, surroundings can be wiped with a sanitizing solution made by adding a teaspoon of bleach to a quart of water. Wipe surfaces and wait ten minutes before rinsing with clean water.

This discussion certainly is not intended to scare anyone away from eating cantaloupe — a fruit that is a good source of the antioxidants beta-carotene and vitamin C. Rather, the goal is to highlight the need for awareness about microbial contamination, the biggest concern when it comes to the safety of our food supply. One bite of contaminated food can have deadly consequences.

Obviously, a cantaloupe can kill, but cure? A case can be made for one particular cantaloupe purchased in 1941 by Mary Hunt, a bacteriologist working at the Agricultural Research Lab in Peoria, Illinois. Thirteen years earlier, Alexander Fleming had discovered the antibiotic properties of a mold that had accidentally drifted into one of his bacterial cultures, and within ten years Florey and Chain had identified penicillin as the bactericidal ingredient. The search was now on to find a mold that would produce a higher yield of the substance.

Researchers throughout the world were asked to send samples of moldy fruit, grains, and vegetables to Peoria for testing. Mary Hunt also took up the challenge, and on her usual shopping trips scoured produce for mold. One day she found a moldy Texas cantaloupe that aroused her interest and brought it to the lab. After cutting out the mold, she and fellow workers enjoyed the sweet taste of the historic fruit. When the mold that had contaminated the melon was steeped in a vat of corn liquor, it yielded twenty times more penicillin than any other mold tested! Within a year, enough penicillin was produced to treat the vast number of World War II battlefield infections, saving thousands of lives. Moldy Mary, as she came to be known, had chanced upon the most celebrated cantaloupe ever grown.

CONFRONTATION WITH *E. COLI* IS A NASTY BUSINESS

You have probably never heard of André Jaccard, but if you eat meat, you have likely benefited from his invention, although some would argue that the term "benefited" has to be qualified. What cannot be argued is that back in the 1970s Jaccard revolutionized an industry by patenting his meat-tenderizing machine.

Tough meat is a tough sell. And what makes for tough meat? An abundance of collagen, the robust protein that makes up what is generally referred to as "connective tissue." To make meat more tender, collagen has to be disrupted either chemically or physically. Moist cooking for a long time will do this as will aging, marinating in an acid solution, or treatment with a plant enzyme such as papain, extracted from papaya. But collagen can also be degraded by grinding, pounding, or "jaccarding."

André Jaccard's invention was a machine that tenderizes meat by piercing it with a series of needles and razor-sharp blades that surgically shred the connective tissue and thereby, at least according to the manufacturer's claim, make any cut of meat "butter tender." Jaccarding also allows more complete penetration of marinades and reduces shrinkage and cooking times. It is easy to see why such mechanically tenderized meat appeals to suppliers, retailers, caterers, and restaurants. After all, it means being able to satisfy palates with cheaper cuts. But it may also mean exposing diners to some nasty microbes, such as the notorious *E. coli* O157:H7.

This wicked bacterium was first identified in 1982 after contaminated hamburgers caused an outbreak of severe bloody diarrhea. It hit the big time in 1993 when undercooked burgers from a fast food restaurant caused the death of four children and sickened 600 other people. That's because if the internal temperature does not reach at least 70°C (158°F), the bacteria can survive and release their toxin.

How is it that we haven't heard of this particular strain of *E. coli* until recently? Actually, that isn't so surprising given that bacteria are constantly evolving as they struggle to survive in a changing environment. This is what appears to have happened to *E. coli*, most strains of which are harmless and commonly inhabit the intestines of humans and animals. But somewhere along the line a particularly vicious strain evolved in the gut of ruminant livestock, such as cattle, deer, goats, and sheep. Curiously, the bacteria do not affect the host in any detrimental way. People, however, can become severely ill should they be infected with the microbe shed by animals in their feces. That's just what we witnessed in fall 2012 when tainted beef from the XL Foods plant in Alberta precipitated the largest beef recall in Canadian history.

Roughly half of all cattle shed *E. coli* O157:H7 in their feces and then end up contaminating their hides as they romp through the muck in feedlots. Then, when the hides are stripped off after slaughter, the bacteria can be transferred to the meat. Similar transfer can occur through removal of bacteria-tainted entrails. Should the contaminated meat then be ground, the bacteria can become distributed throughout.

But not everyone who became sick from meat that originated in the XL Foods plant ate hamburgers — some apparently became ill after eating roasts or steaks. This caused suspicion to be cast on jaccarded meat, given that the process can drive bacteria from surface deep into the tissues, where they may survive, especially if the meat is consumed rare. Meat that has been tenderized in this fashion is not easy to identify, since the holes made by piercing seal up and vanish. If jaccarded cuts were labeled, as is being considered, consumers would at least be alerted to making sure that an internal temperature of 63°C is reached. Of course, there would still be no way of knowing whether meat consumed in restaurants, hotels, or catered events was jaccarded.

And should you think that giving up meat offers protection from the ravages of *E. coli* O157:H7, think again. Tragically, a young girl succumbed to the effects of this bacterium, having been infected by planting a kiss on the cheek of her ailing grandfather who had been stricken after eating tainted beef in a veterans' hall. The little girl lost her life to a hamburger she had never eaten! Another youngster was luckier, surviving a shutdown of her kidneys after being contaminated by *E. coli* O157:H7 at a petting zoo. Long-term consequences are, however, still possible, since the bacterium can also damage the pancreas, lungs, and liver.

Staying away from contact with animals is not a guarantee against contamination either. Actually, there are more outbreaks of *E. coli* O157:H7 infection caused by produce or water than by meat. The major outbreak in 2000 in Walkerton, Ontario, that resulted in seven deaths and 2,000 people becoming sick was traced to manure from nearby farms polluting the water supply. We have seen outbreaks caused by spinach, unpasteurized apple juice, and sprouts. In Europe in 2011, more than 4,000 people became sick from eating fenugreek sprouts. It seems the seeds used for sprouting had been contaminated, probably by exposure to manure. Since manure is commonly used as fertilizer, any fruit or vegetable that has been exposed, particularly if these are eaten raw, is a hazard. Obviously, it is really important to wash all produce well, paying particular attention to sprouts since these are grown under conditions well suited to the growth of bacteria.

And we haven't even talked about fish in China that are partly fed with feces from pigs and geese, shrimp farmers in Vietnam who use ice made from bacteria-laden water, or grape tomato pickers in San Juan who wipe their hands on their pants after answering nature's call in the field. Still worried about smart meters giving off dangerous radiation, artificial sweeteners, or trace chemicals leaching out of water bottles?

SMOKED MEAT

I'll admit it: I like smoked meat. And nobody needs to tell me that it is not good for me. I don't think that every morsel of food that slides down the esophagus has to be evaluated in terms of its nutritional quality. I assure you that it is possible to indulge in this delicacy — not every day, mind you — and still maintain a healthy diet. Just as it is possible to totally shun smoked meat and have a disastrous dietary regimen. True, smoked meat lovers would lose a nutritional debate to the bean sprout and brown rice warriors. But can the delight of biting into a well-stacked smoked meat sandwich be matched by slurping miso soup or chomping on tofu burgers? Not as far as I'm concerned.

Montreal is the smoked meat capital of the world. Period. I would venture to say, though, that most natives have no idea about how this famous product is made. It all starts with beef bellies from Alberta, which, in local lingo, we call briskets. The process of converting these to smoked meat begins by treating the briskets with Chile saltpeter. And here a little history lesson is appropriate.

Perhaps the oldest of all food preservation techniques is salting. Our ancestors discovered that treating meat liberally with salt slowed down the putrefaction process. Salt serves as a dehydrating agent, sucking water out of bacteria and destroying them. But over the years it became apparent that some forms of salt resulted in a better product in terms of keeping qualities, color, and taste!

The reason was a natural contaminant of sodium chloride, namely sodium nitrate, or saltpeter. Today we understand how it works. Microbes in the meat convert nitrate to nitrite. Nitrate is a very effective antibacterial agent, especially against the potentially deadly *Clostridium botulinum* bacterium. It also reacts with myoglobin, a compound found in muscle tissue, to produce the appealing pink color of nitrosyl myoglobin. And, last but not least, sodium nitrite adds a characteristic cure flavor to the meat. Unfortunately, it also adds a health concern. Nitrites can lead to the formation of nitrosamines, which in animals have been shown to be carcinogenic. But more about that later.

The nitrate, which eventually yields nitrite, is dissolved in water and is injected into the meat by means of a specialized machine. Then comes the critical step in terms of flavor. The surface of the meat is rubbed with a blend of "secret" spices. There is salt in the mix, of course. In the old days, it used to come in large grains called corns, hence the expression "corned beef." Coriander, black pepper, chili powder, and bay leaves also add their flavor. Then there is freshly ground garlic! Not only is it an essential contributor to flavor, but also the sulfur compounds it contains have been shown to reduce nitrosamine formation. Finally, the meat is packed into barrels, then cured in a fridge for two weeks.

Now for the all-important smoking process — except that it isn't really smoking. In the old days, meat used to be hung in a smokehouse, exposed to all the compounds generated by burning wood. This cooked the meat, added flavor, and also preserved the meat. Chemicals in smoke, such as formaldehyde, are highly toxic to bacteria. That, of course, is why formaldehyde is used to preserve tissue specimens in the laboratory. Unfortunately, wood smoke also contains a number of compounds that are known to be carcinogens. So how can we smoke meat without worrying about these substances? The truth is that today there isn't all that much worry about the wood smoke because smoking is commonly done in a gas-fired oven, where the meat cooks by convection and the only smoke it is exposed to is generated by the fat that drips down from the meat and burns. This smoking process is still not free of concerns, since the high temperatures generate heterocyclic aromatic amines (HRAS), which are carcinogenic. But there may be a way around this problem, too. Just wait! Some commercial smoked meats are not smoked at all but are injected with smoke flavoring. The less said about these, the better.

After about four hours in the oven, and once the meat has reached an internal temperature of 185°C (365°F), it is removed and sprayed with cold water to stop the cooking process. At this point, it is either vacuum packed or placed in a refrigerator. Prior to eating, the meat has to be steamed for about an hour and a half to restore the water that has been lost during the smoking process. Then it is ready to be cut. And that is a job that requires special training. In Montreal, a "smoked meat cutter" is a highly respected professional, trained to cut against the grain of the meat to produce perfect slices.

Those slices may be perfect visually, but not nutritionally. There's the nagging matter of those nitrosamines, which can disrupt our DNA molecules and initiate nasty processes — perhaps even cancer. But studies have shown that chemicals in tomato juice, such as coumaric or chlorogenic acids, can inhibit this reaction. Research has also shown that vitamin C prevents the reaction of nitrites with amines in the food, or indeed in our bodies. Therefore, an appetizer of tomato juice is great, and orange juice would seem to be the best beverage to accompany a smoked meat sandwich. Purists will surely rebel, claiming that anything other than a black cherry drink is sacrilegious.

Now, what about those heterocyclic aromatics that are byproducts of the cooking process? These form in cooked meat in amounts proportional to the temperature and cooking time. But you know what? Tea contains polyphenols, which have been shown to reduce the mutagenicity of these heterocyclics. Similar compounds are also found in apples. So why not cap off the meal with tea and an apple?

I guess you've gathered by now that there is a moral in here somewhere. Individual foods should not be vilified or sanctified. It is the combination of substances that we put into our mouth that determines our nutritional status. Indeed, smoked meat may not be the most nutritious food. But the nutritional concerns associated with it can be greatly reduced if it is harmonized with other foods and beverages. Unfortunately, pickles and French fries are not the most harmonious accompaniments. Not scientifically speaking, anyway. And please, New

Yorkers, spare me the mail about the wonders of your pastrami and corned beef. I've had both. I've been to the famous delis. They may pile the meat sky high, and it isn't bad, but it isn't "smoked meat."

THE SECRET LIFE OF BAGELS

You should have seen the face of the guy behind the counter in the Manhattan bagel shop when I asked for the smallest, thinnest bagel they had. In a country where excess rules, where the credo is "bigger is better," my request must have come as a shock. But I really needed that thin bagel to save a lecture I was about to give at Columbia University.

The focus of my lecture was on some interesting everyday applications of chemistry, and I wanted to start with a demonstration of how acrylic plastics can make our lives less risky. Dr. Mark Smith, head of emergency at George Washington University Medical Center, had made headlines across America by going public about a "great underreported injury of our times": cuts resulting from bagel slicing. Anyone who has ever risked a mangled hand trying to slice a bagel in half knows exactly what I'm talking about. Luckily, inventors have risen to the challenge and have come up with a variety of devices to ensure that a perfectly good bagel isn't ruined by splattered blood. I had even found one that I really liked. It was a clear acrylic box that held a bagel snugly and had slits down two sides to guide a knife. Not only does it prevent injuries, it also protects bagel lovers from another great scourge — a smoke-filled kitchen. This is what happens when the bigger half of an unevenly sliced bagel refuses to pop up after we've squeezed it into a toaster slot that is too small.

My proposed demonstration of scientific bagel cutting obviously required a victim, and I planned to order that victim at breakfast. Alas, what they brought me was a gigantic roll with a hole in it that looked more like a life preserver than a bagel. I realized that I had a problem. There was no way this thing would fit into my bagel cutter. That's when I ran to the bagel shop and made my unusual request. No shortage of bagels here, but all were as obese as my original. And then, as I stood there frustrated, the door to the back of the shop flew open, and I caught a glimpse of what was going on. Employees were sending raw bagels through a steaming machine. They weren't boiling them; they were steaming them. That's when I decided that New Yorkers didn't need to learn about acrylics. They needed to learn about bagel-making.

Montreal is the center of the bagel world, because here we do it right. For just 180 calories and virtually no fat, you get splendid flavor, unique texture, and a dose of history. According to legend, in 1683, King John III Sobieski of Poland helped save Vienna from Turkish invaders. A grateful Viennese baker created a stirrup-shaped roll to commemorate the bravery of the Polish soldiers. In a German dialect, this roll came to be called *beugel* — meaning "ring," or "bracelet" — because of the large hole in its middle. *Beygel* was the Yiddish version of the name, and from this it was only a short hop to bagel. The bagel was introduced to North America by Jewish immigrants about a century ago, and in Montreal, some of their descendants are still delighting customers by producing bagels in the traditional fashion. There's nothing like the smell and taste of a fresh bagel straight out of the oven. Try the bagel challenge. I defy anyone to buy a dozen and still have a dozen by the time they arrive home. Cannot be done. Not even by someone reared on sliced white bread.

To make this gustatory and health marvel, you don't start with just any flour; you use a flour that is rich in two proteins, glutenin and gliadin. These long, coiled, tangled molecules unfold and line up in long strands when kneaded with water. They also forge cross-links with each other, building a network of proteins known as gluten, which gives dough the elasticity it needs to rise as yeast generates carbon dioxide gas. The baker adds a small amount of sugar to the dough to serve as food for the yeast, along with a little egg for color and flavor. Kneading is critical, because it creates the air pockets into which the carbon dioxide will expand. These air cells will contribute greatly to the final texture. Furthermore, oxygen in the air, introduced during kneading, strengthens the gluten by promoting a chemical reaction that forms sulfur-sulfur links between adjacent protein molecules.

What makes a bagel a bagel, however, is neither the flour nor the kneading. It is the immersion of the hand-formed rings of dough in boiling water prior to baking. Starch molecules in flour are coiled together in tiny granules, but hot water penetrates the granules and causes them to swell. Then the swollen granules muscle their way into, and strengthen, the molecular scaffolding created by the gluten proteins. A classic chewy bagel is the result. Furthermore, the boiling water is not just any boiling water. The baker must dissolve a little honey in it. That's because, in the heat of the oven, sugars in the honey combine with proteins in the dough to form the shiny brown crust prized by bagelites.

Ah, the oven. You can't make a proper bagel without a wood-burning oven. The smoke enhances the flavor, and the burning wood provides just the right temperature. During baking, gluten coagulates, and starch completes its gelatinization. If the temperature is too low, the dough will expand as the volume of the trapped gases increases, but it will then collapse because the gluten and starch have not set. If the oven is too hot, the setting takes place too soon, and the dough does not gain enough volume. It's a touchy business that needs an expert hand. A Montreal hand.

What I saw in New York was not a pretty sight. I saw dough being steamed instead of boiled. I saw electric ovens. I saw jalapeno peppers, chocolate chips, and — believe it or not — bacon bits added to bagels. But even this sacrilege did not prepare me for what I was to see in the frozen food section of the supermarket into which I dashed, hoping against hope, to find a bagel that looked like a bagel. Staring me in the face was the UnHoley Bagel. It looked like a hamburger bun filled with cream cheese. No hole. No class.

By this time, I was getting desperate. I was frustrated by bagels that had no holes and others that were like king-sized donuts with rigor mortis. I had one last chance — Zabar's, Manhattan's most famous food store. No proper bagels here, either, but Zabar's did have something to save the day. An adjustable bagel cutter. It was polyethylene, not acrylic, but I just adjusted my talk accordingly. Thank goodness for American ingenuity.

OH, THAT GLUTEN!

You will never see Novak Djokovic's picture on a box of Wheaties. Djokovic is a super tennis player and is easily in the same league as the athletes who have adorned the Wheaties box since 1934, when Lou Gehrig first urged us to try the "Breakfast of Champions": "There's nothing better than a big bowl of Wheaties with plenty of milk or cream and sugar." Djokovic would disagree. No Wheaties for this champion. Diagnosed as "gluten intolerant" by his nutritionist, Djokovic has given up all foods that contain gluten, the mixture of proteins found mostly in wheat, barley, and rye. He claims that he feels "fresher, sharper, and more energetic."

There is a condition in which gluten plays a critical role. It is called celiac disease. Dr. Samuel Gee of Britain was the first to provide a clinical description of the disease in 1888. He painted a disturbing picture of young children with bloated stomachs, chronic diarrhea, and stunted growth. Dr. Gee thought that the condition may have a dietary connection and put his young patients, for some strange reason, on a regimen of oyster juice. This proved to be useless. Willem K. Dicke, a Dutch physician, finally got on the right track when he made an astute observation during the Second World War. The German army had tried to starve the Dutch into submission by blocking shipments of food to Holland, including that of wheat. Potatoes and locally grown vegetables became staples, even among hospitalized patients. Dicke now noted that his celiac patients improved dramatically! Moreover, in the absence of wheat and grain flours, no new cases of celiac were seen.

By 1950, he had figured out what was going on. Gluten, a water-insoluble protein found in wheat, was the problem. As later research showed, celiac patients' immune system mistakes a particular component of gluten, namely gliadin, for a dangerous invader and mounts an antibody attack against it. This triggers the release of molecules called cytokines, which in turn wreak havoc with the tiny finger-like projections, the villi, that line the surface of the small intestine. The villi are critical in providing the large surface area needed for the absorption of nutrients from the intestine into the bloodstream.

In celiac disease, the villi become inflamed and markedly shortened, effectively reducing their rate of nutrient absorption. This has several consequences. Non-absorbed food components have to be eliminated, and this often results in diarrhea. Bloating can also occur when bacteria in the gut metabolize some of these components and produce gas. But of course, the greatest worry is loss of nutrients. Protein, fat, iron, calcium, and vitamin absorption can drop dramatically and result in weight loss and a plethora of complications. Luckily, if the disease is recognized, and a gluten-free diet begun, patients can lead a normal life.

Diagnosis of celiac disease involves taking a biopsy sample from the duodenum, the uppermost section of the small intestine, via a gastroscope passed down through the mouth. Microscopic analysis shows the damaged villi. Recently, blood tests have also become available. A commonly available one tests for the presence of "anti-gliadin antibodies," but it is not foolproof. Only about half the patients with positive results actually show damaged villi upon biopsy. The anti-tissue transglutaminase test (anti-tTG) is a much better diagnostic tool, but it is available only in specialized labs.

There is a great deal of interest in these tests because of their potential value in identifying celiac cases and perhaps even in screening the population. Celiac disease, which has a genetic component, does not necessarily begin immediately after gluten is first introduced into the diet. The onset of the disease can occur at any age. In adults, the symptoms are usually much less dramatic than in young children. The first signs often are unexplained weight loss and anemia due to poor iron and folic acid absorption. Stools tend to be light colored, smelly, and bulky because of unabsorbed fat. Symptoms can include a blister-like rash, joint and bone pain, stomach ache, tingling sensations, and even headaches and dizziness. Identification of celiac patients is important not only because much of the misery can be prevented by a gluten-free diet but because a recent study showed that over a thirty-year period the death rate among celiac patients was double that expected in the general population.

Risk was increased with increasing delay in diagnosis and poor compliance with diet. The major cause of death was non-Hodgkin's lymphoma, a type of cancer known to be associated with celiac disease. A less severe but more common complication than cancer is osteoporosis, secondary to poor absorption of calcium and vitamin D.

Unfortunately, a gluten-free diet is not that easy to follow. Wheat and barley crop up in a wide assortment of products. Patients have to become veritable sleuths and learn that foods as diverse as ice cream, luncheon meats, ketchup, chocolate, and even communion wafers can contain gluten. Luckily, the Celiac Association has excellent information on dietary dos and don'ts, and a large assortment of gluten-free products based on rice, corn, and soy are now commercially available.

There is no question that people with celiac disease feel better if they scrupulously avoid gluten, but there are also some people who make the same claim despite having no signs of celiac disease. They are estimated to make up around 7 percent of the population and are said to have "non-celiac gluten sensitivity." There is no test for this condition other than feeling better by avoiding gluten. Most of the evidence is anecdotal and similar improvements in health are described by people who avoid artificial sweeteners, shun MSG, eat only raw foods, engage in autourine therapy, or walk barefoot to soak up the earth's energy.

Novak Djokovic did not avoid gluten until his "nutritionist" carried out an "applied kinesiology" test. He asked Novak to stretch out his right arm while placing his left hand on his stomach. He then pushed down on the tennis champion's right arm and told him to resist the pressure which he was able to do. Next, Djokovic was asked to hold a slice of bread against his stomach with his left hand while the nutritionist again tried to push down on his outstretched right arm. This time he was able to push it down easily. The demonstration, Djokovic was told, showed that he was "gluten intolerant," which is why he had suffered so many mid-match collapses in his career.

This "applied kinesiology" test is often used by "alternative" practitioners to diagnose allergies and nutritional deficiencies as well as to promote the sale of "energizing" bracelets. It has zero scientific validity, but that doesn't mean that Djokovic doesn't suffer from non-celiac gluten sensitivity. The correlation with the test may be accidental, but the condition may be real. Djokovic is convinced that avoiding gluten is a factor in his improved play and is not bashful about recommending that everyone give "gluten-free" a shot. And he is not alone. Others who sing the praises of a gluten-free lifestyle include scientific icons such as Gwenyth Paltrow, Miley Cyrus, Lady Gaga, Russell Crowe, and Bill Clinton.

And then there is Dr. William Davis, whose book *Wheat Belly* paints a picture of modern wheat as a satanic grain responsible for diabetes, high cholesterol, osteoporosis, cataracts, wrinkles, rashes, neuropathies, vitiligo, hair loss, and schizophrenia along with "man breasts," "bagel butt," and of course, "wheat belly." If you are scientifically minded, it is worthwhile to read this book just to see how masterfully Davis blends cherry-picked data, inflammatory hyperbole, misused science, irrelevant references, and opinion masquerading as fact into a recipe for a cure-all. Some of the "science" is just absurd. He talks about how wheat DNA has been mutated by exposure to sodium azide, and then points out that "the poison control people will tell you that if someone accidentally ingests sodium azide, you shouldn't try to resuscitate the person because you could die, too, giving CPR." The fact that sodium azide is a toxic chemical has nothing to do with its use in inducing mutations in genes. There is no azide in the product and inducing mutations to achieve beneficial traits is a standard technique used by agronomists.

Davis's argument for wheat causing osteoporosis is equally bizarre. He describes how wheat can give rise to sulfuric acid when it is metabolized. This is indeed correct. One of the amino acids in wheat protein, cysteine, does end up releasing some sulfuric acid in the body. And the body does use phosphates from bone to neutralize excess acid. The amount of acid released into the bloodstream from wheat is trivial, yet Davis calls it an "overwhelmingly potent acid that rapidly overcomes the neutralizing effects of alkaline bases." Poppycock. (Appropriately that term originates from the Dutch term for "soft dung.")

That, though, isn't the worst of it. Davis panics readers with totally irrelevant statements about sulfuric acid causing burns if spilled on the skin. Get it in your eyes and you will go blind. True, but what does that have to do with traces formed in the blood from cysteine? Sulfuric acid in acid rain erodes monuments, kills trees and plants, Davis informs us. Yes, it does. But linking this to eating wheat is an example of mental erosion. Davis also claims that proteins in wheat break down to peptides that have opiate-like activity and lead to wheat addiction. If that were true, we had better avoid spinach, soybeans, meat, dairy, and rice because these also contain the same protein fragments.

Davis also claims substantial weight loss by avoiding wheat. "If three people lost eight pounds, big deal," he says. "But we're seeing hundreds of thousands of people losing 30, 80, 150 pounds." Really? Where is this documented? It isn't surprising, though, that some people do lose weight on the Wheat Belly diet given that cutting out wheat products results in a reduced caloric intake.

There is also accumulating evidence that improvements in health by avoiding gluten have nothing to do with gluten but rather with "fermentable oligosaccharides, disaccharides, monosaccharides, and polyols" dubbed FODMAPs. These wheat components are poorly absorbed and travel through to the colon, where they provide a scrumptious meal for the bacteria that live there. The problem is that these bacteria produce copious amounts of gas that distend the gut and cause pain as they dine on the FODMAPs. Unfortunately, other foods, including many fruits and vegetables, also contain these troublesome sugars, so a low FODMAP diet is difficult to follow.

In the meantime, Novak Djokovic is winning titles and is winning other athletes over with his gluten-free diet. Wouldn't it be interesting to see how he would perform if somebody managed to sneak some gluten into his food? I also wonder how Lou Gehrig would have done had he traded in his Wheaties for Rice Chex or Corn Flakes. I suspect just as well.

AN ODE TO THE OAT

I'd like to take a look at Papa Bear's blood test. His triglycerides are probably high from slurping all that honey, but his cholesterol level is likely to be just fine, thanks to his love of porridge. In fact, all the members of the Bear family, with their penchant for oats, can serve as nutritional role models. I, for one, am following in their footsteps. And I'm managing to keep pace with science.

The Scots got this one right. Porridge is one of their staples. Scotch oats are steeped not only in water and milk, but also in a good dose of tradition. I understand that the mush must be stirred clockwise, with the right hand, using a spurtle, which is a sort of wooden stick especially made for this purpose. And the porridge is to be eaten from a birch-wood bowl. "Porridge sticks to the stomach and scrubs out the bowels," the Scots maintain. True enough. Oats really do have a high satiety value. Essentially, this means that they take a long time to digest and therefore keep you feeling full longer. Indeed, in a study comparing oatmeal to cornflakes as breakfast foods, researchers found that subjects who ate oatmeal consumed one third fewer calories for lunch. So, oats can help you lose weight.

The bowel-scrubbing bit makes sense too. In more ways than one. Oats contain fiber. Fiber is the structural part of plants, grains, fruits, and vegetables; it cannot be broken down by enzymes in our digestive tract and therefore cannot provide nutrition. In other words, most of what you eat turns into you, but fiber passes through. There are two kinds of fiber: insoluble and soluble. Cellulose is the classic insoluble fiber, whereas pectin, found in fruits, is an example of the soluble variety. The former keeps us regular, reduces the risk of diverticulitis, and helps eliminate substances that may play a role in colon cancer. But it is beta-glucan, the soluble fiber in oats, that is causing a stir. Solid research has shown that while oats produce no nutritional miracles (no single food does), those who consume them regularly can experience health benefits: lower blood cholesterol levels, a decrease in high blood pressure, healthier arteries, and better diabetes control.

Some of this information about oats is not new. Just think back to the oat bran craze. Retailers couldn't keep the stuff on the shelves. Rumors of a new shipment sent anxious shoppers rushing to the supermarket, only to have their hopes dashed when they found that the booty had already been snapped up. Why was there such a feverish interest in a product traditionally considered animal feed, not human food? Because some tantalizing studies showed that oat bran, the outer covering of the grain, is an excellent source of soluble fiber, which has the ability to reduce cholesterol. Some researchers offered a theory to explain how this happens. Beta-glucan absorbs water in the intestine and forms a viscous slurry that traps cholesterol from food as well as some of the bile acids needed for digestion. Since these compounds are made in the body from cholesterol, their removal from the digestive tract forces the body to synthesize more. The result is a depletion of the cholesterol in the blood. Good stuff. But there was a problem. The public never got the proper message about how much oat bran they would have to consume to impact their blood cholesterol levels. And this was no small amount.

To reduce blood cholesterol by roughly 5 percent, a person needs to eat three to four grams of beta-glucan a day. More is not better. At higher doses, one experiences a sense of fullness; bloating and gas production become apparent. Now, a 5 percent reduction doesn't sound like a lot, but it can lower the risk of a heart attack by as much as 10 percent. We can find this amount of beta-glucan in one cup of cooked oat bran, or one and a half cups of oatmeal. Three packets of instant oatmeal will do it too. But oat bran cookies, oat bran chips, and oat bran gum will not. Yet manufacturers flooded the market with these silly products, hoping to capitalize on the oat bran mania. The products had no effect on cholesterol, and they tasted lousy to boot. Little wonder the oat bran fad faded quickly. Too bad. Because, when consumed in the right quantities, oats really do deliver the goods. They can do more than just lower cholesterol — they can reduce blood pressure.

A pilot study in Minnesota focused on a group of patients who took at least one medication for hypertension. Researchers asked half of them to consume about five grams of soluble fiber per day in the form of one and a half cups of oatmeal and an Oat Square (an oat-based snack); they asked the other half to eat cereal and snacks with little soluble fiber. Oat consumption reduced blood pressure in these patients significantly. Indeed, about 50 percent of them were able to give up their medication. How oats lower blood pressure is not clear, but it probably has to do with modifying insulin response. The pancreas secretes insulin, which enables our cells to absorb glucose from the bloodstream after a meal. A glucose surge triggers a quick insulin response, but if such surges are frequent, insulin becomes less effective, and the body needs to produce more and more. This leads to a condition known as insulin resistance. Researchers suspect that such insulin resistance plays a significant role in elevating blood pressure by constricting blood vessels. Soluble fiber slows the absorption of nutrients from the gut and blunts the insulin response. This also explains why oats can help diabetics control their blood sugar levels.

And if that weren't enough to boost your appetite for oats, just consider that oats contain a unique blend of antioxidants, including the avenanthramides, which prevent LDL cholesterol (the bad cholesterol) from being converted to the oxidized form that damages arteries. So, it isn't hard to see why I've become a real oat fan. And I've become an even bigger fan since I discovered steel-cut oats. These are oat grains cut into thirds but not rolled into little flakes by a machine. They have a great nutty flavor. Admittedly, they do take longer to cook and require constant stirring. That's why I'm searching for a good spurtle. If you've got one, I'll trade you for my oat soup recipe.

Ah, heck. I'll give you the recipe anyway. Bring twelve cups of chicken stock to a boil. Add six sliced carrots, three sliced parsley roots, one cup of peas, one cup of diced onion, two tablespoons of canola oil, two tablespoons of soy sauce, two mashed garlic cloves, and two cups of rolled oats. Simmer for forty minutes and add salt and pepper to taste. I bet even Baby Bear would love it.

I'M A BREAD AND PROPIONATES MAN

I'll admit it. I like bread. Not any bread of course since I'm very aware of the problems linked with the overconsumption of refined carbohydrates. I am not the least bit enticed by the spongy, tasteless, pale, packaged loaf that masquerades as bread in many a supermarket, but give me a slice of Austrian schinken-brot smeared with a bit of butter, and I'm in heaven. That's right, butter. In moderate amounts it can fit into a healthy diet. And to me, it sure tastes better than margarine.

Certainly schinkenbrot is not one of your more common bakery items. So I really wasn't surprised to see a fellow shopper eyeing the sample in my cart with some suspicion. The bread quickly proved a catalyst for conversation, and I explained that I was partial to this loaf not only for its flavor, but also for its nutritional value. It's basically made of whole rye kernels, whole wheat flour, rolled oats, and sourdough culture, ingredients that are surely superior to the refined white flour and sugar that dominate the classic American "toast" bread.

Intrigued, she picked up my schinkenbrot and began to peruse the ingredients. I saw her brow furrow as she read the list, and with an air of incredulity, she lifted her eyes and pierced me with a look that I thought would have been reserved for someone who had committed a capital crime. "You're not really going to eat this, are you?" she sputtered with considerable bewilderment. After confirming that I had not been planning a fun-filled afternoon of pigeon feeding in the park and that the bread was indeed destined for the dinner table, I queried her concern about my diet. Her reply? "It says right here, 'May contain calcium propionate!'" She blurted out the term "calcium propionate" as if it were synonymous with "poison," which alerted me to what this interaction was all about: another rampant case of chemophobia.

I thanked the shopper for bringing this issue to my attention because I had never noticed the word "may" on the label of my preferred bread. I don't like the notion of "may contain." I would prefer a guarantee that the bread does contain calcium propionate. You see, I'm not a great fan of moldy food. Molds not only make for unsightly green splotches, but some produce decidedly dangerous compounds, which is why we add preservatives such as calcium propionate to foods. They prevent the growth of molds while allowing yeast to flourish — an obvious advantage when it comes to baking bread. And that's not all that calcium propionate does. It also inhibits the formation of rope in bread. The spores of certain bacteria, such as *Bacillus mesentericus*, are often present in flour and germinate under the moist, warm conditions needed to make bread rise. These bacteria are not harmful to humans, but they change the texture of the dough and produce sticky yellow stringy patches that make for an unpalatable bread. Propionates prevent this undesirable result.

Are propionates safe to eat? They are. These compounds cruise through our bodies all the time, and they don't have to be introduced through bread. Bacteria in our intestine feed on fiber, the indigestible part of fruits, vegetables, and grains, and convert it into a variety of compounds that include propionic acid. This acid is then absorbed into the bloodstream. Far from being harmful, some studies have shown that such short-chain fatty acids can reduce the risk of colon cancer and may even be helpful in preventing other diseases of the digestive tract.

Propionates, as derivatives of propionic acid are called, also occur naturally in our food supply. Perhaps the best example is Swiss cheese. The texture and flavor of this cheese is due to the addition of a starter culture that includes the bacterial species known as *Propionibacterium shermanii*. These bacteria break down some of the fat to produce carbon dioxide gas, which explains the presence of holes in the cheese. They also produce propionic acid, which is responsible for some of the cheese's characteristic nutty flavor. Swiss cheese contains roughly 1 percent propionates by weight, far more than the amount used as a preservative in bread.

So I am absolutely untroubled by the presence of calcium propionate in my schinkenbrot. And I like the taste and the whole grain ingredients. I'm comforted by a Finnish study that showed that elderly men who ate just three slices of old-fashioned, fiber-rich rye bread per day reduced their risk of fatal heart attacks by 17 percent. (We're not talking about the type of rye bread sold in most bakeries here; that version is made mostly from refined flour.) And if you are interested in weight control, you may find comfort in an Australian study for which researchers fed volunteers seven different kinds of bread and rated their "satiety" value after two hours. Soft white bread scored the lowest. Coarse-textured, high-fiber breads ranked the highest. Subjects who ate the latter consumed fewer calories over the rest of the day.

Is there any problem with eating high-fiber dark breads like my schinkenbrot? Maybe a little one. These dark breads are often referred to as pumpernickel, from the German words *pumpern*, meaning "to break wind," and *nickel*, meaning "devil." Pumpernickel was thought to be so hard to digest that its victims would pass wind like the devil. I am not privy to Lucifer's dietary habits, so I do not really know what it means to pass wind like the devil. But it may not be a bad thing. Passing gas is a sign of high fiber intake, which has all sorts of health benefits. Perhaps I should have mentioned this fact to my grocery-aisle acquaintance when I encountered her again in the checkout line. She could have used the info because I noted a loaf of "organic" white bread in her carriage, whose package proudly declared,

"No preservatives added." Oh well. Maybe she likes mold. Or maybe she just prefers days with no wind.

AGITATE FOR ICE CREAM

Nancy Johnson of Philadelphia had a problem. She loved ice cream. But she found that making it was a struggle. She'd often spend up to an hour shaking the metal pot containing her mixture of cream and sugar before the stuff would freeze. And all that time she had to keep the pot immersed in a bath of ice and salt. There had to be a better way. So, in 1843, Nancy dreamed up the ice cream maker. She placed a metal can filled with ingredients in a wooden bucket and packed it with layers of ice and salt. Then she attached a hand crank to a brace positioned across the top of the bucket and ingeniously connected it to a paddle that would churn the mix as it froze. Thanks to Nancy, anyone could now make ice cream at home.

The concept of making ice cream is simple enough. Take some cream, add sugar and flavor, and freeze the mixture. Pure water freezes at $0°C$ ($32°F$), but by dissolving any substance in water we lower its freezing point. So the ice cream mix, with all of its dissolved sugar, requires a temperature lower than zero to solidify. Now picture what happens if we place this mix in a container and then immerse it in a bucket packed with ice. The original temperature of the ice is well below zero (just check the temperature in your freezer), but the surfaces that are in contact with the air will quickly warm up to zero degrees and begin to melt. The water from the melted ice will also be at zero degrees, and this mixture of ice and water will remain at that temperature as long as any ice is present. But at zero degrees, the ice cream mix will not freeze. However, if we sprinkle salt on the ice, we create a different scenario. As before, the surface of the ice warms up and melts. The water dissolves the salt, and soon the pieces of ice are swimming in salt water. Since this liquid has a lower freezing point than pure water, the ice will lower its temperature until the new freezing point is reached. In other words, the ice cream container is now surrounded by salt water, which is at a temperature well below zero. The mix freezes.

But just freezing the mix won't give you ice cream. It will yield a dense, solid mass filled with ice crystals. Hardly mouth-watering stuff. If you want good taste, you must agitate. Shaking or mixing the ingredients during the freezing process is the key to making good ice cream. This accomplishes two things. First, it minimizes the size of the ice crystals that form; second, it blends air into the ice cream. The smaller the ice crystals, the smoother the ice cream. But it is the pockets of air blended into the product, known as the overrun, that give it its prized foamy consistency. Nancy Johnson's hand-cranked device minimized crystal formation and incorporated air admirably. Indeed, ice cream manufacture today still uses the same principle.

Human ingenuity does come to the fore when ice cream makers are unavailable. During World War II, American airmen stationed in Britain and pining for ice cream discovered that the gunner's compartment in a bomber had just the right temperature and vibration level for making the sweet treat. They would put the ingredients into a can before a mission, stow it in the gunner's compartment, and then look forward to returning to base with a batch of freshly made ice cream.

The method of simultaneous mixing and freezing solves the main problems of ice cream manufacture, but it does introduce a complication. Cream essentially consists of tiny fat globules suspended in water. These globules do not coalesce, because each is surrounded by a protein membrane that attracts water, and the water keeps the globules apart. Stirring breaks the protein membrane, the fat particles come together, and the cream rises to the top. This effect may be desirable when we're making butter but not when we're making ice cream. There is a simple solution: we can add an emulsifier to the mix. Emulsifiers are molecules that take the place of the protein membrane, since one end dissolves in fat and the other in water. Lecithin, found in egg yolk, is an excellent example. That's why even the simplest ice cream recipe requires some egg yolk.

There is nothing like freshly made ice cream. Its smooth, airy consistency affords us a break from reality; it's a throwback to childhood and a less complex world. Storing ice cream, however, also presents a problem — the dreaded heat shock. By taking the container out of the freezer, for example, we may cause the surface of the ice cream to melt. When it refreezes, it forms larger ice crystals, resulting in the crunchy texture that so terrifies ice cream lovers. The commercial solution? Add some wood pulp.

Now, don't get all worried — we're not talking about adding sawdust to ice cream. Microcrystalline cellulose is a highly purified wood derivative that sops up the water as ice cream melts and prevents it from refreezing into crystals. Cellulose is indigestible, so it comes out in the wash, so to speak. Guar gum, locust bean gum, or carrageenan, all from plant sources, can also be used for the same purpose. Although lecithin is a good emulsifier, there are others that are more commercially viable. Mono- and diglycerides or polysorbates disperse the fat globules very effectively.

For those of you yearning for homemade ice cream but unwilling to deal with the salt and ice, here's a solution. Find a chemist friend who can provide you with some liquid nitrogen and supervise your activity. Place the mix in a Styrofoam container, add liquid nitrogen, and stir. The mix freezes almost immediately and develops just the right foamy consistency as the nitrogen evaporates. With a little practice, you can outdo Nancy Johnson.

A final word of warning, though. Ice cream may be addictive. A study conducted at the U.S. Institute of Drug Abuse suggests that eating it stimulates the same receptors in the brain as certain drugs. If you run into this problem, you may want to sample one of the new flavors that commercial manufacturers are tinkering with. Garlic, spinach, pumpkin, or tuna ice cream is sure to curb your craving.

TURKISH ICE CREAM IS
BAD NEWS FOR ORCHIDS

Who would have thought that a love of ice cream could be driving some varieties of orchids towards extinction? Not any ice cream, mind you, but a special variety favored mostly in Turkey, called *dondurma*, which in Turkish means "freezing." Dondurma is more resistant to melting than other ice cream and has a thicker, stretchy consistency as well as a characteristic flavor. The texture and flavor are mostly due to the inclusion of the hardened sap of the mastic tree, which grows mostly on the Greek Island of Chios. The sap hardens into a resin composed of a complex mixture of simple carbohydrates and polysaccharides that can impart an elastic texture when pressure is applied artificially or by chewing. Indeed, the name "mastic" derives from the Greek verb for "gnashing the teeth," which also is the source of our English word, "masticate." Mastic resin also contains a variety of compounds in the terpene family that provide a unique pine-like flavor and aroma. Besides ice cream, the resin is also added to puddings, sauces, Turkish delight, perfumes and body lotions. It is a relatively expensive commodity due to the rarity of the mastic trees.

The exudate of the mastic tree has a history of use as a medicine as well. The ancient Greek physicians Dioscorides, Hippocrates, and Galen all mention the medicinal properties of mastic, mostly for colds and digestive problems. While an effect on colds is questionable, there may actually be something to the digestive issues. A few studies have shown that compounds in mastic have antibacterial properties and can improve peptic ulcers by killing *Helicobacter pylori* bacteria. They may also reduce tooth decay by diminishing the population of oral microbes that secrete acids. There are also stories about mastic extracts being used as breath fresheners in the harems of sultans in the Middle Ages. Whether the resin was chewed by the sultan or his ladies isn't clear.

The resistance to melting is mostly due to the addition of salep, flour made from the roots of a genus of orchid. Its main component is glucomannan, a form of dietary fiber composed of glucose and mannose molecules joined in long chains. It has a remarkable ability to absorb water, which means it can keep ice cream firm even when it is melting. This is also the reason that glucomannan has been marketed as a diet aid. Unlike digestible carbohydrates such as starch, glucomannan is resistant to breakdown by our salivary or pancreatic enzymes. As a dose of indigestible glucomannan sits in the stomach or small bowel before passing on to the colon, it absorbs a great deal of water. This bulky mélange of water and fiber makes for a feeling of fullness and curbs the appetite. A few short-term studies have shown more efficient weight loss on a low-calorie diet that incorporates about 4 grams of glucomannan per day. However, a number of companies have hyped glucomannan supplements, mostly extracted from the tubers of the konjac plant, way beyond what research has shown.

Like mastic resin, salep has a long history of folkloric use. Sixteenth-century philosopher and physician Paracelsus believed that nature bestowed specific shapes on plants to mark their curative benefits. This eventually became known as the "doctrine of signatures," based on German theologian Jakob Böhme's claim that God had marked objects with a "signature" to give clues about their purpose. Parts of plants or animals that resembled human body parts were thought to have a beneficial effect on those parts. Paracelsus suggested that orchid roots when ingested would restore a man's virility and passion since they resembled the male testes. Soft on evidence.

In the seventeenth century, beverages made from salep became popular in England, possibly because of the aphrodisiac connotation. Often, however, British orchid roots, known as dogstones, were substituted for salep. While the consumption of salep in England has faded, in Turkey, drinks and ice cream made from orchid flour are increasing in popularity. This is bad news for orchids since it takes a couple of thousand tubers to produce a kilo of salep. Orchids are difficult to cultivate on farms and the main supply is wild orchids. Harvesting these has raised the specter of local extinctions, not only in Turkey, but in Africa and Iran as well, with millions of tubers being illegally harvested for export to Turkey. This is disturbing, especially given that glucomannan is available from konjac root, and indeed, in Turkey producers are turning to this instead of using orchids. Still, the cachet of the traditional process is hard to overcome. Researchers are working on ways to culture orchid seeds in the lab with hopes of developing seedlings that can be commercially cultivated. That then may lick the Turkish ice cream problem.

SAY "CHEESE"

Cheese producers were cheesed off. People were just not eating enough veal. Slaughterhouses were running short of calf stomachs and the cheese industry was feeling the pinch. There was not enough rennet to meet the demands of turophiles (that is, cheese lovers; the Greek word *turo* means "cheese") around the world.

Rennet, you see, is critical to the cheese-making process. At least it is if you want to indulge in something that is a little more exciting than cottage cheese. Traditionally, rennet has been made by washing, drying, macerating, and brining the lining taken from the fourth stomach of calves. This process leads to a product that is a mixture of two enzymes, chymosin and bovine pepsin, both of which can coagulate milk and convert it into cheese. Why does the stomach lining of mammals contain these enzymes? Because they are needed for proper digestion. If milk did not coagulate to some extent in the stomach, it would flow through the digestive tract too quickly, and its proteins would not be sufficiently broken down into absorbable amino acids.

Enzymes are specialized protein molecules that serve as biological catalysts. They make possible the myriad chemical reactions that go on inside our bodies all the time. Specifically, chymosin and bovine pepsin are proteases, which means that they catalyze the breakdown of proteins, a task that is central to the milk coagulation process. Milk consists of about 87 percent water, 5 percent lactose, 3.5 percent fat, 3.5 percent protein, and 1 percent minerals. The protein content consists mostly of casein molecules, which are insoluble in water and aggregate into tiny spheres called micelles. Since their density is comparable to the surrounding solution, micelles remain suspended. Actually, there are three kinds of casein molecules: alpha-casein, beta-casein, and kappa-casein. Within the micelle, the alpha- and beta-caseins are curled up like a ball of string and are held together by kappa-casein, which functions much like an elastic band. The job of chymosin is to break the band and allow the casein molecules to stretch out and form a long, tangled network of protein molecules that settles out of the solution. Fats and minerals get snared in this protein net and — presto! We have cheese!

Chymosin is the ideal enzyme for catalyzing this process. In an acidic environment it snips kappa-casein specifically, allowing the other caseins to unwind. In the stomach, cells that secrete hydrochloric acid create the acidic environment, whereas in cheese-making, a bacterial starter culture that converts lactose into lactic acid is added before the rennet. Bovine pepsin is not quite as suitable as chymosin because it has a more general protease activity, snipping caseins in a variety of ways. This enzymatic action weakens the protein network needed to trap fat and results in a lower yield of cheese. Furthermore, some of the protein fragments it produces have a bitter taste and subtly alter a cheese's flavor profile.

By the 1960s, the shortage of rennet was becoming critical. The stomachs of older animals were pressed into service, but the resulting rennet was not really suitable. As an animal ages, chymosin production decreases and pepsin production increases. So scientists had to step in and take the bull by the horns, as it were. Actually, instead of bull horns, they grabbed chicken bones. Researchers at the University of Guelph in Canada discovered that rennet enzymes would bind very well to porous chicken bones and milk could be pumped through a matrix of these bones to start the curdling process. With this procedure, the same amount of rennet would go a lot further than if it were just dumped into vats of milk. While an interesting possibility, the method was never commercialized because a number of companies found that by using a technique called anion exchange chromatography, they were able to separate pure chymosin from the stomach extracts of older animals. This technique made 100 percent pure chymosin available for the first time, but the process was complicated and not cheap. However, there was another way to compensate for the lack of calf rennet.

In the 1960s, researchers had discovered that certain fungi, *Rhizomucor miehei* being a prime example, were capable of producing enzymes that would cleave proteins in much the same way as chymosin. This discovery meant that cheese could be produced without using any animal rennet at all. The breakthrough not only addressed the rennet shortage, but it also made possible the production of cheese that met the needs of vegetarians. Such cheese was also kosher because there was no mixing of milk and meat during production. But purists claimed the taste was not the same. They may well have been right. Fungal enzymes have greater proteolytic, or protein-breaking, activity than chymosin and can give rise to "off" flavors.

Then genetic engineering entered the picture and essentially solved the chymosin problem. The bit of DNA, the gene, that gives the instructions for the formation of chymosin was isolated from calf cells and copied, or cloned. It was then successfully inserted into the genetic machinery of certain bacteria (*E. coli*), yeasts (*Kluyveromyces lactis*), and fungi (*Aspergillus niger*), which dutifully churned out pure chymosin. Approved in 1990 by the U.S. Food and Drug Administration, chymosin became the first product of genetic engineering in our food supply. It is 100 percent identical to the chymosin found in calf stomach, but because it does not come from an animal, it is acceptable to consumers who do not want meat products in their cheese.

Extraordinary precautions were taken before chymosin, made by recombinant DNA technology, was marketed. Regulators ensured that no toxins of any kind had been introduced and that no live recombinant organisms were present. Indeed, the product contained nothing but pure chymosin. Cheese made with it is completely indistinguishable from that produced with animal rennet. In any case, chymosin itself is degraded during cheese-making and none is left in the finished product. Today in North America, over 80 percent of all cheese is made with chymosin produced by recombinant DNA technology. Cheese-makers no longer have to worry about a shortage of calf stomachs, and turophiles can satisfy their critical taste buds. Thanks to biotechnology, they can say "cheese" and smile.

CHEW ON THIS

I'll tell you up front that I don't like chewing gum. I'm familiar with the studies that have shown chewing may reduce tooth decay, help with weight management, and even reduce stress, but I'm not won over. And it isn't because I'm worried about the "carcinogens, petroleum derivatives, embalming fluid ingredients, or chemicals that cause diarrhea or mess with your digestive system." I don't buy these accusations that permeate the Internet, authored by scientific luminaries with self-conferred titles such as The Food Babe.

My aversion to gum probably traces back to elementary school when one of my teachers had a unique punishment for anyone caught chewing gum in class. The criminal had to climb on a chair and recite: "The gum-chewing student and the cud-chewing cow differ somehow. I know, it must be the intelligent look on the face of the cow!" Ever since witnessing such a "sentence" being carried out, I can't look at a masticator without comparison to a cow, decidedly to the animal's advantage. Memory sure is a mysterious thing. And therein lies a gummy story to chew on.

Back in 2002, researchers at Northumbria University in England assigned seventy-five subjects aged twenty-four to twenty-six to either chew gum, mimic chewing without gum, or not chew at all while performing both short- and long-term memory tests. Gum chewers scored significantly higher. Although the robustness of this study has been criticized, it did unleash speculation about why chewing gum may aid memory. For one, research has shown that chewing gum increases blood flow to the brain and activates the frontal and temporal cortex, probably by enhanced oxygen transport. Since these regions are known to play a role in cognitive function, increased memory seems a possibility. Another option is context-dependent memory, implying that information is more easily recalled in an environment similar to the one experienced while learning, particularly if a smell is involved. For example, students studying while exposed to the scent of chocolate perform better when exposed to the same scent as they write exams.

In 2011, Dr. Matthew Davidson of Stanford University's School of Medicine explored the memory enhancement effect further, adding another twist. He fortified the gum with substances that have been associated with improved cognition. Many studies have suggested improved alertness with caffeine, so it was a natural additive. In fact, the U.S. army has introduced caffeinated gum in military rations. Davidson also added an extract of the *Ginkgo biloba* tree along with vinpocetine derived from the lesser periwinkle, both of which have been shown to enhance blood flow to the brain. Also included was an extract of a creeping herb known as *Bacopa monnieri*, which in at least one placebo-controlled, double-blind study was shown to improve learning rate and memory. Rosemary and peppermint were also added, mostly to take advantage of their strong scent that may enhance recall. The sixty-two participants were divided into three groups; they chewed either the "Think Gum," ordinary bubble gum, or nothing while engaged in learning as well as during recall.

The results were not exactly spectacular but the herbal-caffeinated gum chewers did perform better on most tests than the non-chewers, in some cases by as much as 30 percent. Bubble gum chewers did only marginally better than non-chewers, implying that the effects noted were due to the gum's additives rather than to chewing. While Think Gum may be a helpful study aid, the product's advertising slogan of "stop cheating, start chewing" is a bit hard to swallow.

What is even harder to swallow is the diatribe directed at chewing gum on some websites and circulating emails. Here is one gem: "Gum is typically the most toxic product in super-markets and is likely to kill any pet that eats it . . . loaded with harmful ingredients and chemicals." (Hmm, I wonder what sort of ingredients don't contain chemicals?) One of the "harmful ingredients" highlighted is BHT (butylated hydroxy-toluene), a preservative. The basis of the scare? It is found in embalming fluid! Yes it is, but so what? BHT is an approved food additive that prevents fatty substances from reacting with oxygen. It is actually sold in pill form in health food stores as a dietary antioxidant.

Another "toxin" we're supposed to worry about is vinyl acetate. This is indeed a worrisome substance if exposure is sig-nificant, but that isn't the case with gum. These days, gum base is mostly made of synthetic rubber rather than natural substances like the sap of the sapodilla tree, commonly known as chicle. The synthetics include styrene-butadiene rubber, polyethylene, and polyvinyl acetate (PVA). In theory, PVA may contain traces of vinyl acetate, the chemical from which it is made, but the amount is trivial. The devil, as they say, is in the details.

The scares about gum that go around today, however, can take a back seat to the one that rocked the Egyptian city of Mansoura in 1996, when loudspeakers blared out warnings to young girls about the evils of chewing gum "laced with aphrodisiacs capable of transporting the most innocent female into a sexual frenzy." Sold under the brand names Aroma and Splay, the gum was said to be an Israeli attempt to corrupt Egyptian youth. Laboratory analysis by Egypt's Ministry of Health found nothing other than the usual ingredients, but that didn't stop authorities in Mansoura from closing any kiosk that dared sell the gum that supposedly caused young ladies to engage in immodest activities.

Talking about immodest activities, how about a gum that claims to increase bust size? Bust Up gum contains miroestrol and deoxymiroestrol, two estrogen mimics found in the underground tubers of a plant called *Pueraria mirifica*. Lack of evidence for the reputed benefit notwithstanding, it is curious that some women who may be concerned about trace amounts of estrogen mimics leaching out of plastics will swallow significant amounts of phytoestrogens when it comes to trying to improve their appearance.

And now for a final bit of nonsense. Calcium casein peptone-calcium phosphate (CCP) is an ingredient in Trident gum. Various alarmists warn about it with inspired comments like "casein is a milk-extractive that was linked with the Chinese baby formula poisonings." Casein is indeed a protein found in milk, but it has nothing to do with the poisonings that were due to formula being adulterated with melamine in an attempt to increase protein content. CCP actually remineralizes tooth enamel by delivering calcium and phosphate beneath the tooth's surface. The scare about it is just another example of the rampantly galloping chemophobia we're witnessing. That stresses me. Maybe I should chew gum after all.

THINKING ABOUT COCONUT OIL

Think about this: what has no mass, doesn't occupy space, has no mobility, cannot be touched, and yet exists? A thought! And what a mysterious thing it is! Just about all we know for sure is that it is created in the brain and that there is an energy requirement to generate it.

Whenever we think, the brain "burns" more glucose, which is its main fuel supply. It stands to reason that any sort of inhibition of this glucose metabolism can have a profound effect on brain function. We know, for example, that a rapid drop in blood glucose, as can be precipitated by an overdose of insulin, quickly causes a deterioration in cognitive performance. This is because so much glucose is absorbed by muscle cells that little is left for the brain.

Alzheimer's disease is characterized by a progressive decline in the rate of glucose metabolism in the brain. This impaired use of glucose is paralleled by a decline in scores on cognitive tests. Exactly why glucose use is affected in Alzheimer's is not clear. It may be a function of the buildup of amyloid protein deposits that are the hallmark of the disease, although it is also possible that the deposits are not the cause, but are rather the result, of impaired metabolism. Recently there have even been questions about whether amyloid deposits actually play a significant role in Alzheimer's disease. In any case, improving the brain's ability to generate energy in the face of low glucose metabolism seems a worthy avenue to explore.

The most obvious approach would be to supplement the diet with glucose and provide sufficient insulin for its absorption into cells. But insulin cannot easily be delivered specifically to the brain, and its systemic administration can cause problems in other tissues. So is there another option? A clue can be found in studies of people who are experiencing starvation. When there is a lack of glucose available from the diet, the body tries to meet the brain's demand for energy by tapping its abundant stores of body fat.

Fat, however, cannot be used directly as fuel; it first has to be converted to smaller molecules called ketone bodies. The buildup of these in the bloodstream results in ketosis, a condition that is not encountered when there is an adequate intake of carbohydrates, the source of glucose. It can, however, occur in diabetes when an insulin shortage prevents glucose absorption into cells, which then have to resort to the use of ketone bodies to supply energy. That's why acetone, a ketone body, appears in the breath of diabetics who fail to administer their insulin properly. Ketosis can also be encountered when low-carbohydrate regimens such as the keto, paleo, or Atkins diets are followed. It is the breakdown of fat to yield ketone bodies that results in weight loss.

Now back to Alzheimer's disease. An extremely low-carbohydrate diet can conceivably increase ketone bodies delivered to the brain, but such diets are difficult for many to follow and long-term consequences are unknown. But there may be another approach. It turns out that not all forms of dietary fat are handled by the body the same way. So-called long-chain fatty acids, having at least thirteen carbons in the chain, as found mostly in animal products, are readily stored by the body, whereas the "medium-chain fatty acids" that contain six to twelve carbons tend to be metabolized in the liver to ketone bodies. This presents a potential therapeutic application for Alzheimer's disease. Why not just supplement the diet with medium-chain fatty acids? They're not hard to find. You don't have to look further than coconut oil.

At least one published trial lends support to the idea. Patients in the early stages of Alzheimer's disease showed an improvement in cognitive performance tests administered ninety minutes after treatment with a single 40-gram dose of medium-chain fatty acids. This finding flew pretty well under the public radar until pediatrician Dr. Mary Newport's story started to circulate on the Internet. Actually, it was her husband's story that got people talking. Steve Newport was diagnosed with Alzheimer's and was fading quickly. His wife did what most people do these days: she let her fingers do the walking on the keyboard. As a physician, she knew that the drug treatments available were not very effective and became intrigued when she came across the research that had linked medium-chain fats to increased metabolism in brain cells. What was to lose by giving her husband a couple of tablespoonfuls of coconut oil every day?

The very next day, Steve was scheduled for a routine cognition test and showed a surprising improvement over his previous performances. Dr. Newport obviously decided to continue the regimen and reports that after two months her husband was once more reading avidly, resumed jogging, and even started to do volunteer work at a hospital. But should he miss his morning oil, he quickly becomes confused and experiences tremors. Swallowing the regular dose brings quick improvement.

So what are we to make of all this? Is there a cure for Alzheimer's disease that is being ignored by conventional medicine? Not likely. But that is not to say there isn't something to the medium-chain fatty acid story. However, it is a little disturbing that the source for the Internet buzz is an article written by Dr. Frank Shallenberger. Let's just say this good doctor is not a candidate for a staff position at Harvard Medical School. Following multiple disciplinary actions for gross incompetence, he surrendered his California license and moved to Nevada, where he later pleaded guilty to another count of medical malpractice. He now writes a newsletter about "real cures," such as "ozone therapy," and pushes medium-chain fatty acids for Alzheimer's disease.

What we have here is one interesting study published in the literature that in no way shows reversal of Alzheimer's, an intriguing personal account that begs for independent verification, and some overly optimistic statements from a physician who has had disciplinary actions against him for incompetence.

But let's not throw the baby out with the bathwater. The theory behind boosting levels of ketones such as acetone to enhance cellular energy production in the brain has merit and needs further exploration. Don't think, though, that drinking nail polish remover is the way to go. But as far as coconut oil goes, as we so often say, more research is needed.

WHAT'S IN A VITAMIN NAME?

You've got to love the plot. A meteorite falls to earth and begins to ooze a revolting goo that dines on humans. It makes its way into a movie theater and, in a classic scene, surges from the projection booth, ready to gobble up everyone in its path. Audiences in 1958 were absolutely terrified by *The Blob*, which was destined to become a science fiction classic. The film also introduced actor Steve McQueen to the world and made me a fan. I watched his films and followed his career until his unfortunate death from lung cancer in 1980. Steve introduced me to some great movies and also to vitamin B17, which he had been taking at a Mexican cancer clinic in an attempt to halt the progress of his disease.

By 1980, I had been teaching for a while and had developed several lectures on vitamins. But none of the books or publications I consulted had ever referred to vitamin B17. And for good reason, as it turned out. Vitamins are substances that must be included in the diet in order to maintain health and prevent certain deficiency diseases; they cannot be synthesized by the body. The first vitamin-deficiency disease to be recognized was scurvy, described as early as 1550 B.C. by the Egyptians in the *Ebers Papyrus*. In the sixteenth and seventeenth centuries, when long ocean voyages became common, thousands of sailors died from scurvy, which is characterized by spongy gums, loose teeth, and bleeding into the skin and mucous membranes. The first clue that scurvy was a diet-related disease came from Indigenous North American peoples who showed French explorer Jacques Cartier that a brew made from pine needles could cure the condition. In 1747, James Lind, the Edinburgh-born naval surgeon, discovered that eating oranges and lemons prevented scurvy, but it took another fifty years before the British navy required sailing vessels to carry supplies of lemons or limes. Around the same time, British Captain James Cook discovered that fresh fruits and sauerkraut also prevented scurvy. Finally, in the 1930s, Hungarian scientist Albert Szent-Györgyi isolated the scurvy-protective factor and named it "vitamin C." Why? Because the idea of naming vitamins by letters had already been introduced some twenty years earlier and A and B were taken.

The letter designation for vitamins goes back to the early part of the twentieth century. When the mechanized rice mill was introduced in Asia, a new disease that came to be called beriberi appeared. *Beriberi* means "weakness" in the native language of Sri Lanka and describes a condition of progressive muscular degeneration, heart irregularities, and emaciation. Kanehiro Takaki, a Japanese medical officer, studied the high incidence of the disease among sailors in the Japanese navy from 1878 to 1883. He discovered that on a ship of 276 men, where the diet was mostly polished (hulled) rice, 169 cases of beriberi developed and twenty-five men died during a nine-month period. On another ship, there were no deaths and only fourteen cases of the disease. The difference was that the men on the second ship were fed more meat, milk, and vegetables. Takaki thought that the few deaths on the second ship had something to do with the protein content of the sailors' diet, but he was wrong.

About fifteen years later, a Dutch physician in the East Indies, Christiaan Eijkman, noted that chickens fed mostly polished rice also contracted beriberi but recovered when fed rice polishings (the hulls). He erroneously thought that the starch in the polished rice was toxic to the nerves. Finally, Casimir Funk, a Polish chemist, showed that an extract of rice hulls prevented beriberi. He thought that this substance fell into the chemical category of amines, and since it was vital to life he called it "vitamine." When the substance turned out not to be an amine, the final e was dropped.

A short time later, E.V. McCollum and Marguerite Davis at the University of Wisconsin discovered that rats given lard as their only source of fat failed to grow and developed eye problems. When butterfat or an ether extract of egg yolk was added to their diet, growth resumed, and the eye condition was corrected. McCollum suggested that whatever was present in the ether extract be called fat soluble A, and that the water extract Funk had used to prevent beriberi be called water-soluble factor B. When the water-soluble extract was found to be a mixture of compounds, its components were given designations with numerical subscripts. The specific antiberiberi factor was eventually called vitamin B1, or thiamine. These vitamins had a common function. They formed part of the various enzyme systems needed to metabolize proteins, carbohydrates, and fats. Some of the compounds in Funk's water extract eventually turned out to offer no protection against any specific disease and their names had to be removed from the list of vitamins. As other water-soluble substances required by the body were discovered, they were added to the B-vitamin list.

Other vitamins were subsequently identified and given the designations D and E in order of their discovery. Vitamin K was so called because its discoverer, the Danish biochemist Henrik Dam, proposed the term "koagulations vitamin" because it promoted blood coagulation. Are there still unrecognized vitamins? Not likely. Patients have now been kept alive for many years through total parenteral nutrition (TPN), which involves using an intravenous formula that incorporates the known vitamins. No nutritional-deficiency diseases have shown up in spite of the fact that no vitamin B17 is to be found in the formula.

So what is vitamin B17? Essentially, a scam. In the 1950s, Dr. Ernst T. Krebs came up with the idea that a compound extracted from apricot pits, called amygdalin, was able to selectively target cancer cells and destroy them by releasing cyanide. Krebs and his son, Ernst Jr., became the first proponents of administering this compound, an approach called Laetrile therapy. When the government began to ask for evidence that the drug worked, Krebs changed his approach. The public was becoming familiar with the benefits of vitamins, so he decided to convert Laetrile into one. Krebs then claimed, without any evidence, that cancer is caused by a deficiency of vitamin B17. Numerous studies carried out since have failed to show that this substance can treat or prevent any type of cancer. Whatever amygdalin may be, it is not a vitamin.

And what happened in the finale of *The Blob*? The authorities eventually figured out that the menacing goo couldn't stand the cold. So the Air Force found a way to transport it to the Arctic and put it in a deep freeze. Which is precisely what should be done to the unsubstantiated claims being made about vitamin B17.

MAN CANNOT LIVE ON CORN ALONE

Italian cuisine is one of my favorites. Except for polenta. I have never developed a taste for that odd corn mush, which was once a dietary staple of poor Italians. When explorers returned home to Europe from North America with corn, it quickly became popular with landowners because of its abundant yield. These landowners often paid the farm workers they hired to grow the corn with a share of the crop, and corn became an important part of their diet.

By the late 1700s, however, it was becoming evident that the sharecroppers who subsisted on corn were an unhealthy bunch. One could easily recognize them by their crusty, reddened skin. Pellagra, from the Italian for "rough skin," became a common term for the condition. Most people believed that it was caused by eating spoiled corn. Rough skin was not the only symptom the poor sharecroppers had to worry about. The disease was often characterized by a red tongue, a sore mouth, diarrhea, and dementia — before it killed its unfortunate victim. Pellagra came to be referred to as the disease of the four Ds: dermatitis, diarrhea, dementia, and death.

By the early 1900s, pellagra had reached epidemic proportions in the southern U.S. It ravaged the poor, especially cotton pickers. Some sort of a communicable infection now seemed a more probable cause than contaminated corn. Rupert Blue, the U.S. Surgeon General, stepped in and assigned his top epidemiologist, Dr. Joseph Goldberger, to solve the mystery of pellagra. Many of the pellagra victims ended up in insane asylums, so these institutions seemed appropriate places to start the investigation. Goldberger soon realized that while many inmates had the symptoms of pellagra, no doctor, nurse, or attendant showed signs of the disease. He noted the same phenomenon in orphanages, where children often developed pellagra but staff members never did. This was inconceivable if pellagra were an infectious disease. So Goldberger began to ponder the lifestyle differences between the asylum and orphanage inhabitants and the attending staff of these institutions. He also began to speculate about differences in diet.

Goldberger observed some pretty dramatic differences. Both inmates and staff got plenty of food, but the variety was not the same. While the attendants dined on milk, butter, eggs, and meat, the pellagra sufferers had to subsist mostly on corn grits, corn mush, and syrup. Goldberger suspected that some sort of dietary deficiency might be triggering pellagra. But he uncovered one troublesome finding. In one orphanage he studied, most of the younger children showed symptoms of pellagra, but the older ones seemed to fare much better. This mystery was solved when Goldberger discovered that the resourceful older children were supplementing their diet with food that they snitched from the kitchen.

It was obvious to Goldberger what the next step in his investigation had to be. He must obtain government funding to add meat and dairy products to the diets of the orphans and the asylum inmates. He did so, and the results were miraculous. Almost all of the pellagra victims regained their health. But if he was to prove the dietary connection conclusively, Goldberger would have to conduct one more critical experiment. He would have to show that pellagra could be induced by a faulty diet. And where was he going to find volunteers for such a study? In prison. Convicts would do anything to get out of jail. So

Goldberger approached the director of the Rankin State Prison Farm in Mississippi and outlined his idea. The director agreed to cooperate. He would release any prisoner who volunteered to take part in Goldberger's study upon the study's completion.

The volunteers were soon lining up to lend Goldberger a hand, especially after the doctor explained the protocol. To the prisoners, it sounded like a cakewalk. For six months, they could eat to their heart's content, as long as they confined themselves to a menu of corn biscuits, corn mush, corn bread, collard greens and coffee. Then they would be freed. After about five months, though, the fun went out of the experiment. The convicts began to suffer from stomach aches, red tongues, and skin lesions. Goldberger had proven his point. Unfortunately, he did not have a chance to cure his patients, since, true to his word, he'd had them released. The convicts quickly scattered, wanting no more of Goldberger's dietary schemes.

It would seem that the problem of pellagra was solved. But many scientists who had pet theories about contagion remained unconvinced. In a letter to his wife, a frustrated Goldberger described these colleagues as "blind, selfish, jealous, prejudiced asses." He would show them that pellagra was not a contagious disease! Dr. Goldberger organized a series of "filth parties," at which he swallowed and injected himself, his wife, and supportive colleagues with preparations made from the blood, sputum, urine, and feces of pellagra patients. Nobody came down with the disease. Goldberger had made his point by eating excrement.

Unfortunately, Dr. Goldberger did not live to see the day when the "pellagra-preventive factor" was finally identified. In 1937, scientists put the finger on niacin, one of the B vitamins. Corn, as it turns out, is a very poor source of niacin; when people — like Goldberger's inmates and orphans and convicts — eat little else, they develop pellagra, a deficiency disease. It's a shame that the convicts dispersed before the doctor could arrange to follow them up. It would have been interesting to see how they eventually fared, whether they suffered strokes or age-related macular degeneration, a leading cause of visual impairment. Why? Because recent studies demonstrate that lutein, a pigment abundant in corn, may be protective against both of these conditions. Apparently, lutein concentrates in the eye and protects it from the harmful effects of blue light.

Ultrasound measurements of the thickness of carotid arteries, a predisposing factor for stroke, reveal an inverse correlation with blood levels of lutein. Furthermore, lutein-fed mice that were genetically engineered to develop atherosclerosis developed lesions only half as large as those seen in mice on normal feed. Sounds pretty good. It almost makes polenta sound appealing.

LESSONS FROM POPEYE

The most famous landmark in Crystal City, Texas, is a statue of Popeye the sailor man. He's squeezing his trademark can of spinach, ready to save Olive Oyl from the clutches of Bluto. Crystal City, you should know, is the spinach capital of the world. Its citizens erected the statue in 1937 to honor the character who single-handedly boosted spinach consumption and helped save an industry. But there may be more of interest in Crystal City than Popeye's statue. I think someone should look into the incidence of heart disease there. In Crystal City, spinach is a way of life — and, I suspect, a longer one. That's because spinach is an outstanding source of folic acid, a B vitamin that is increasingly being linked with a plethora of health benefits. Let me explain.

Our story starts in the hallowed halls of Harvard University, far from the spinach fields of Crystal City. It was here, in 1969, that Dr. Kilmer McCully became involved in the unusual case of a boy who died at the age of eight from a stroke. The boy had suffered from a rare condition in which a substance called homocysteine builds up in the blood. Homocysteine is a normal metabolite of methionine, a common amino acid found in virtually all dietary proteins. A healthy person's body quickly processes it, but it accumulates in those suffering from homocystinuria, like McCully's young patient. An autopsy clearly revealed the cause of death. The boy's arteries were like those of an old man. Could the damage have been caused by excess homocysteine? McCully wondered. To investigate this further, he needed to examine other children who were afflicted with the same condition.

It didn't take him long to reach a conclusion: children with high homocysteine levels sustain artery damage typical of that seen in older men. And then, to prove his point, McCully injected homocysteine into rabbits, where it caused artery damage. This was enough evidence to suggest a revolutionary idea: homocysteine is a risk factor for heart disease. McCully proposed that high levels of the substance cause damage quickly, while levels that are only slightly elevated take longer to wreak havoc. Excited by his findings, he submitted a paper to *The American Journal of Pathology*. But instead of getting famous, McCully got sacked.

Harvard denied him tenure, supposedly because of his unorthodox theory about heart disease. Members of the medical establishment had declared that cholesterol was the main culprit, and they could see no room for homocysteine in their scenario. Eventually, however, Dr. McCully would be vindicated. And, somewhat fittingly, one of the first studies to show the validity of the homocysteine theory was carried out at the Harvard University School of Public Health. In 1992, researchers reported on an analysis of disease patterns in over 14,000 male physicians. Those subjects whose blood levels of homocysteine ranked in the top 5 percent had a heart attack risk that was three times greater than the risk calculated for subjects with the lowest levels. Numerous other studies have shown a similar relationship. A high homocysteine level (above 12 micromoles per liter) seems to be a clear, independent risk factor for heart disease.

Knowing about a risk factor is not much good unless we can do something about it. And in the case of homocysteine, we can. Let's take a moment to explore the relevant biochemistry. Homocysteine forms through the action of certain enzymes on methionine. Once it has formed, one of two things will happen. It will either be reconverted to methionine or metabolized to glutathione, a powerful antioxidant. Both of these pathways require the presence of B vitamins. The body needs folic acid and vitamin B12 to change homocysteine back to methionine, and it requires vitamin B6 for the glutathione route. You are probably starting to get the picture. Inadequate levels of these B vitamins lead to increased levels of circulating homocysteine, which in turn causes arterial damage and heart disease.

The B-vitamin doses we need to keep homocysteine in check are not extreme. About 400 micrograms of folic acid, 3 micrograms of B12, and 3 milligrams of B6 should do the job. While we can certainly get these through diet, the fact is that many of us don't. Indeed, the average intake of folic acid in North America is about two hundred micrograms — far from adequate. This is where spinach comes in. It is an outstanding source of folic acid, particularly if we eat it raw. So, go for that spinach salad. And may I suggest dressing it with orange juice? Just one cup contains two hundred micrograms of folic acid. You can also throw in some green beans or cooked brown beans, also great sources of folate. You'll be helping your heart, and other parts of your anatomy as well. A study of 25,000 women showed that those who consumed the most folic acid were one-third less likely to develop precancerous polyps in their colon. And if that isn't motivation enough to seek out foods that are rich in folic acid, then consider that it may even lower the risk of Alzheimer's disease. Yup, you heard right.

Researchers at the University of Kentucky explored the Alzheimer's connection because they were aware of the extensive evidence showing that women who took folic acid supplements during pregnancy had babies with fewer neurological birth defects, such as spina bifida. Could folic acid affect the nervous system later in life? the researchers wondered. A group of nuns in Minnesota who had willed their bodies to scientific research provided the answer. Those who had ingested adequate amounts of folic acid throughout their lives were less likely to succumb to Alzheimer's disease. This finding was corroborated by researchers at Tufts University, who fed spinach to rats and found that it not only prevented but also reversed memory loss. It seems that homocysteine can damage nerve cells the same way it damages blood vessels.

What all of this comes down to is that Popeye was right. That's why I'm dismayed by his fading popularity among children. We could use his nutritional support. Especially when you consider that some researchers suggest we could prevent 50,000 heart attacks a year in North America simply by increasing our folic acid intake. That spinach salad with orange dressing is looking mighty good.

DIETARY SUPPLEMENTS:
TO TAKE OR NOT TO TAKE?

Should you take a daily multivitamin? It's a pretty simple question to answer, one would think. After all, there have been literally thousands of studies on how vitamin and mineral intakes relate to health. More than 100 million people in the U.S. and Canada believe the question has been answered and take a variety of daily supplements to protect themselves against disease. Can all these people be wrong? Let's see.

Back in 2002, researchers in Oxford, England, decided to look into the issue of preventing heart disease through vitamin supplementation. Since it is well accepted that LDL cholesterol (the "bad" cholesterol in our blood) is more likely to accumulate in artery walls after undergoing reaction with oxygen, it was reasonable to expect that antioxidants such as vitamins E, C, and beta-carotene would reduce the risk. Over 20,000 adults who had diabetes, high blood pressure, high cholesterol, or were otherwise at risk for heart disease were enrolled in a major study. Half the subjects received a daily supplement of 600 international units (IU) vitamin E, 250 milligrams vitamin C, and 20 milligrams beta-carotene, while the others got a placebo. Blood tests showed that this regimen significantly increased the levels of these nutrients in the blood. And what happened after five years of this regimen? Nothing. There was no difference in any form of disease, or in death rate, between the experimental group and the control group. Of course, one can argue that the subjects were already at high risk and the disease processes were already underway when supplementation was begun, and the supplements were powerless to reverse them. So how about looking at giving supplements to an initially healthy population? That's been done too.

One study enlisted over 83,000 healthy American physicians who filled out questionnaires about their dietary habits. About 30 percent reported that they used daily multivitamin supplements. After five and a half years, just over 1,000 had died from some form of cardiovascular disease. There was no relationship to supplement intake. Doctors who had been popping vitamin pills were as likely to die from heart disease as those who had not. Now, one can argue that these physicians were cognizant of good nutrition, ate a balanced diet, and therefore had no vitamin deficiencies that could be corrected by taking supplements. So what about supplements and average people? A study on this question has also been completed.

When some 600 subjects over the age of sixty were assigned to take 200 IU vitamin E, a multivitamin, both, or a placebo, the results were surprising. Those who took vitamin E had worse colds and respiratory infections over a period of fifteen months. Worrisome? Then how about the study of over 70,000 post-menopausal nurses that showed those who consumed the most vitamin A from foods and supplements over an eighteen-year period had the greatest risk of bone fractures? Ready to flush your vitamins down the toilet? Well, hold on.

A study at Canada's Memorial University examined the effects of giving seniors a daily multivitamin for a year. There was a significant improvement in the subjects' immediate memory, abstract thinking, and problem-solving abilities. An English study of over 20,000 people found that those who had the highest levels of vitamin C in the blood lived the longest. A large American study found that maternal use of vitamin supplements reduced the risk of neuroblastoma, a childhood cancer. It is well established that taking folic acid supplements during pregnancy can reduce neural-tube defects in babies. But that's not all folic acid can do. A statistical review of over

20,000 people in ninety-two studies showed a significant decrease in the risk of heart disease with higher blood levels of folic acid. And there's still more. Women who have higher levels of folate in their blood appear to have greater protection against breast cancer, especially if they consume two or more alcoholic beverages per day, a known risk factor for breast cancer. Reduced blood levels of folic acid are also linked to colorectal cancer and long-term use of supplements has been shown to lower risk by as much as 75 percent. Vitamin D intake also offers protection against breast cancer in some studies. Selenium deficiencies in the diet have been linked to prostate cancer. A randomized, double-blind, placebo-controlled study in North Carolina showed that respiratory tract infections could be dramatically reduced in diabetics who took a combination of vitamins and minerals. And here is something really intriguing: a study published in 2004 reported its findings after having followed almost 5,000 people aged sixty-five and older in Utah. Those who took daily supplements of vitamin E (400 IU) and vitamin C (500 milligrams) reduced their risk of developing Alzheimer's disease by almost 80 percent. Strangely, neither vitamin alone was effective. Ready to rescue those vitamins from the toilet?

The point of reviewing these studies is to show that it is possible to support either side of the supplement issue by selectively looking at the scientific literature. However, scientific conclusions should hinge on an examination of all the evidence. And that is just what researchers do when they establish the recommended daily doses of nutrients. Yes, it is possible to get such amounts from a balanced diet. But surveys show that while North Americans do not suffer from acute vitamin deficiencies, suboptimal intakes are common. About 75 percent of the population consumes inadequate amounts of folic acid. We know from laboratory experiments that low levels of folic acid lead to DNA damage and breaks in chromosomes, both of which are linked to cancer. Many people, especially seniors, fall short of the recommended daily intake of vitamin B12, vitamin C, and zinc. The human body requires some forty micronutrients for effective function and while we may not be sure of the optimal amounts, we do know that a large segment of the population has inadequate intakes.

Many people take a daily multivitamin supplement as "nutritional insurance" although there is no evidence that supplement takers are healthier. Still, there is no harm in taking a daily multivitamin/mineral supplement that has no more than 100 percent of the recommended daily intake of each nutrient. If anyone is inclined to do so (I'm not), look for supplement with no more than 4,000 IU vitamin A (better if some of this comes from beta-carotene, a vitamin A precursor), about 400 IU vitamin D, 90 milligrams vitamin C, at least 2.4 micrograms vitamin B12, 400 micrograms folic acid, 10 milligrams zinc, and no more than 10 milligrams iron (although menstruating women may require up to 18 milligrams). The jury is still out on vitamin E, but up to 400 IU presents no problems. There is no consensus on vitamin C, but 500 milligrams per day is unlikely to cause any side effects. Some companies advertise that their products are "more pure" or "better absorbed." But all vitamins have to meet standards of purity and absorption. And there's no need for any ginseng, lutein, ginkgo, or whatever happens to be currently in vogue to be included. There will be no benefits from the amounts of these compounds put into a multivitamin. Above all, remember that Mae West's famous comment "Too much of a good thing is wonderful" most certainly does not apply to supplements. Her observation that "it takes two to get one into trouble" is far more appropriate.

THE COLD FACTS ABOUT VITAMIN C

Vitamin C and the common cold. You would think that the exact link would be clear by now. After all, it's been decades since Linus Pauling suggested that vitamin C was the answer to this viral misery, and his theory spurred a host of investigations. Had this claim been made by anyone else, the scientific community would have yawned and ignored it. But this was Linus Pauling, a Nobel Prize winner, and perhaps the greatest chemist of the twentieth century. He had contributed so much to our understanding of the chemical bond, the structure of proteins, and the mystery of sickle-cell anemia that many scientists thought his "gut feeling" about vitamin C merited further examination.

Humans are one of just a few species incapable of synthesizing vitamin C, a distinction we share with other primates, guinea pigs, fruit-eating bats, and the red-vented bulbul, a curious bird. Pauling believed that we lost the ability to manufacture the vitamin from components in food somewhere along the evolutionary line, and that we were now paying the penalty. Pauling maintained that the amounts ingested in the diet may be enough to protect us from scurvy, the classic disease brought on by vitamin C deficiency, but that this was not enough for optimal health.

Linus Pauling was one of my chemical heroes when I was growing up, and I remember the excitement I felt the first time I heard him speak at a conference. It was in the early 1970s, and I can still vividly recall the great man strutting around the stage, brandishing a vial holding the amount of vitamin C a goat produces in a day, telling the enraptured audience, "I would trust the biochemistry of a goat over the advice of a doctor." Hmmmm, I thought, that didn't sound very scientific. But this was Linus Pauling. He had to know what he was talking about!

Well, it seems that far more distinguished members of the scientific community than me weren't quite so sure and decided to put the vitamin C hypothesis to a test. Let's mount some clinical trials, they proposed, and check this out. And how appropriate that was! After all, the first real clinical trial in history involved vitamin C. That goes back all the way to 1747, and the pioneering work of a Scottish ship's surgeon, Dr. James Lind, who is usually credited with having discovered that scurvy — a disease that rotted gums, caused joints to become swollen, robbed the body of energy, and often killed its victims — could be cured with citrus fruit juice.

Dr. Lind would have approved of the clinical trials that various researchers organized to check out Pauling's vitamin C/common cold hypothesis. Over sixty controlled studies have examined the effects of vitamin C supplements on the common cold, in some cases using up to several grams a day. You can pick and choose among these studies to "prove" whatever point you want to make. If you want to show no effect whatsoever, an Australian study of 400 volunteers taking various doses of vitamin C over eighteen months is the example of choice. If you want the opposite result, check out an American study of 463 students over two years. And if you want the true bottom line, here it is. The evidence that vitamin C supplements can prevent the common cold is very sketchy, although there may be an effect in people who have an extremely low dietary intake of the vitamin.

This is corroborated by the findings of two researchers, one Australian, the other Finnish, who examined the published studies to date and subjected them to critical analysis. They looked at some fifty-five placebo-controlled trials that used at least 200 milligrams of vitamin C supplements a day. The results are not spectacular. In people who took vitamin C regularly to prevent colds, there were no fewer colds, but there was a slight reduction in the number of days they experienced symptoms: 14 percent for children; 8 percent for adults. Interestingly, there was a 50 percent reduction in the incidence of colds in marathon runners, skiers, and soldiers exposed to significant cold and/or physical stress. There was also some evidence that taking large doses of vitamin C, as much as 8 grams, on the day that cold symptoms appear can shorten the duration of a cold. Indeed, a recent study showed that the synthesis of cytokines, chemicals the body generates to fight viruses, is increased within hours of taking a gram of vitamin C.

Taking vitamin C prior to extreme physical exertion or exposure to cold stress therefore makes sense, as does taking large doses the first day that cold symptoms appear. In any case, the impact of vitamin C supplements on the common cold is not a major one. If it were, we would have seen it conclusively in the studies, and we would not be debating the issue.

FREE RADICALS BAD, ANTIOXIDANTS GOOD: IS THAT SO?

There is one thing we know for sure about antioxidants: they sell products. Unfortunately, that is just about the only thing we know for sure about this fascinating class of chemicals. In the public mind, though, antioxidants are superheroes that wage war against those evil free radicals that conspire to rob us of our health and our youth. And according to a variety of supplement promoters, the antioxidants that are naturally present in our food supply are not enough to protect us from the free radical onslaught. We need reinforcements in the form of whatever pill, capsule, or potion they happen to be pushing. And the concoctions usually come with plenty of testimonials about lives turned around. But what they lack is compelling evidence.

The story usually goes something like this: we need oxygen to live. No contesting that. Any student who has studied glycolysis and the dreaded Krebs cycle will recall the critical role that oxygen plays in the production of cellular energy. Basically, glucose reacts with oxygen to yield carbon dioxide, water, and energy. But there are also some byproducts. These are the notorious free radicals, also referred to as reactive oxygen species, or ROS. And they are reactive. Should they take aim at important biomolecules such as proteins, fats, or nucleic acids, they can wreak molecular havoc.

Our body, however, doesn't just stand by as the electron-deficient free radicals try to satisfy their hunger for electrons by ripping essential molecules apart. It musters its defenses. And those defenses are the antioxidants, a wide array of compounds linked by the ability to neutralize reactive oxygen species. They encompass enzymes such as superoxide dismutase, vitamins A, C, and E, and various polyphenols derived from plant products in our diet. Since plants produce oxygen through photosynthesis, they have had to evolve protective mechanisms to deal with oxidation. It is reasonable to propose that we can benefit from the antioxidants they churn out. So far, so good.

Populations that consume more fruits and vegetables are healthier. That is also more or less correct. But why should this be so? This is where the fly plunges into the ointment. The seductive argument is that produce is loaded with antioxidants and that by scavenging free radicals, these chemicals are responsible for the health benefits. But fruits and vegetables are also loaded with all sorts of compounds that have biological activity unrelated to free radical scavenging. Flavonols in cocoa beans, for example, dilate blood vessels by triggering the formation of the messenger molecule nitric oxide. Isoflavones in soy interact with estrogen receptors. Curcumin in turmeric inhibits an enzyme that catalyzes the formation of proinflammatory prostaglandins. Salicylic acid in apples has an anticoagulant effect. In spite of all this, the focus has been on antioxidants. A simple formula emerged: free radicals are bad, antioxidants are good. And the marketers ran with that one. Fast enough to blur the facts.

Any substance that showed antioxidant potential in the laboratory was elevated to the status of a quasi-drug. Shelves sagged under the weight of exotic juices, green tea pills, pine bark extracts, capsules filled with various carotenoids, and, of course, vitamins C and E in every conceivable form. The contest was on for the antioxidant championship of the world. Products vied with each other to claim the highest oxygen radical absorption capacity, or ORAC rating. ORAC is a measure of the ability of a sample to neutralize free radicals in a test tube. But the body is not a giant test tube, and ORAC values do not necessarily translate into biological significance. Many polyphenols that can put on an impressive antioxidant performance in the test tube may not even be absorbed from the digestive tract.

Vitamins C and E, along with beta-carotene, readily decimated free radicals in lab experiments and became the poster boys for antioxidant supplements. But when researchers got around to carrying out clinical trials, the results were disappointing. Most found no benefit. One actually showed an increase in lung cancer risk in smokers taking beta-carotene supplements. Some studies even claimed an increased risk of premature mortality in people who regularly supplemented with antioxidants. What's going on here? Could it be that free radicals are not the villains they have been made out to be? As is so often the case in science, issues that seem straightforward on the surface become more complicated with a little digging. So let's dig a little.

It turns out that white blood cells generate and unleash free radicals in their fight against bacteria and viruses. So clearly, in the right quantities, at the right time, free radicals can be health enhancing. Furthermore, the production of free radicals for such defense purposes is sensed by other cells that then fire up their internal defenses and produce enzymes, such as catalase and superoxide dismutase, that can deal with larger numbers of potentially dangerous free radicals. Sort of like how exposure to a small amount of toxin prompts the system to deal with larger insults.

One interesting theory suggests that antioxidants in food may actually work by generating small doses of health-promoting free radicals. When an antioxidant neutralizes a free radical by donating an electron, it itself becomes a free radical, but a potentially much less damaging one. Still, it may be threatening enough to stimulate the body's own antioxidant defenses. Antioxidant supplements may not work as well as antioxidants found in food because the doses are too high, and they may suppress free radical formation excessively. Sound far-fetched? Well, consider this.

It is well known that exercise improves insulin sensitivity, which in turn helps manage type 2 diabetes. However, exercise also increases the formation of reactive oxygen species as cells "burn" more glucose to generate the energy needed. In a German study, forty healthy young men were given an exercise regimen to follow for four weeks. Half the subjects were asked to take a daily supplement of 1,000 milligrams of vitamin C and 400 IU of vitamin E. Surprisingly, insulin sensitivity improved only in the men not taking the supplements! Furthermore, production of superoxide dismutase and glutathione peroxidase, the enzymes that protect against free radicals, was increased by exercise, but again only in the subjects not taking the supplements. It seems that the free radicals produced by exercise-induced oxidative stress provide the signal for increased insulin sensitivity and for revving up antioxidant defenses. The researchers concluded "supplementation with antioxidants may preclude these health-promoting effects of exercise in humans." So, I think I'll stick to getting my antioxidants from my daily five to seven servings of fruits and vegetables.

Life sure is complicated. There are no simple solutions. As H.L. Mencken said, "For every complex problem there is a solution that is simple, neat, and wrong." And the relationship between diet, health, and supplements is indeed a complex problem.

TO LABEL OR NOT TO LABEL, THAT IS THE QUESTION!

I place "consumer information" on a pretty high pedestal. In fact, these days, I spend much of my time trying to provide reliable scientific information to a public often confused by the deluge of apparently contradictory data. So it should come as no surprise that I'm a strong advocate of product labels that provide us with useful information. But, at the same time, I'm also wary of the ease with which inappropriate labeling can mislead consumers.

Let me begin with an interesting and disturbing case in point. A few years ago, a television infomercial touted the benefits of Rio Hair Naturalizer, a product aimed mostly at Black women wishing to relax their curls. The guest "expert" on the show spoke of the horrors caused by other hair straighteners that relied on "harsh chemicals," and then pointed to the product's label with its "all natural" and "chemical free" claims. The host followed up on this by leading the audience in a chant of "What do we want to be? Chemical free!" Obviously, the product was not chemical free; nothing but a vacuum fits that description. It was, however, natural. The Rio hair relaxer was formulated with cupric chloride, a naturally occurring mineral. It relaxes curls, but also causes hair loss, green discoloration, blisters, and scalp burns. Over 50,000 women filed court complaints before the FDA put a stop to the nonsense by arguing that the label falsely claimed the product was chemical free.

Far less serious than this, but also misleading, are products that proclaim themselves to be "cholesterol free." Technically, the statement may be correct, since cholesterol is found only in animal products, and these foods do not contain any ingredients derived from animal sources. But there are two problems here. First, the insinuation that other similar foods may not be cholesterol free, and second, that being cholesterol free offers a significant health benefit. For example, a vegetable oil that screams "no cholesterol" on its label suggests that other such oils do contain the substance. A popular cookie may declare it contains no cholesterol while it supplies huge amounts of fats that constitute a greater risk for heart disease than dietary cholesterol does. If we are interested in reducing heart disease risk, we should push for labels that tell us not only the fat content per serving but also how this is distributed. It would be great to know the trans fatty acid content as well as the ratio of omega-3 to omega-6 fats. A high ratio here may be protective against a host of diseases. Now that would be useful information.

And it is useful information that we hope to glean from a label, isn't it? That brings us to the thorny issue of labeling genetically modified foods. Suppose a label states "contains no GMOS." What message does that send? Does it not suggest that there is a reason to avoid GMOs? But genetically modified foods on the market have been approved by stringent regulatory agencies both in Canada and the U.S. They have been assessed for their effect on human and animal health as well as environmental safety. Not a single case of human disease has ever been attributed to any such food. But in spite of this, if consumers do want to see such labels, then the information has to be verifiable. This presents a problem: there may not be anything to verify since there is no chemical difference between soy oil made from genetically modified soybeans and oil made from non-GMO beans. We may also see dishonest producers jumping on the bandwagon and labeling everything in sight as "GMO-free," including products for which genetic modification is not an issue. Do we need to see tomatoes labeled as GMO-free when no genetically modified tomatoes are on the market?

What if we go in the other direction and focus on labeling foods that are sourced from genetically modified plants? Clearly, a bag of soybeans, canola, or corn grown with the aid of this technology can be so labeled. But what about a ready-to-eat meal that lists cornstarch as an ingredient? The starch may have come from corn with a gene inserted to produce the insecticidal *Bt* toxin, but the starch cannot be distinguished from starch that comes from non-GM corn. Would it be labeled? What about meat from animals raised on genetically modified feed? Or eggs from chickens similarly raised? How about a tomato modified with a gene from another variety of tomato? Or cheese produced using the enzyme chymosin that was made by recombinant DNA techniques, but is identical to chymosin found in the stomachs of animals? If regulations that require foods containing GM components to be labeled are introduced, what will be the maximum amount allowed to be present without such a label? 1 percent? 5 percent? 0 percent? How can this possibly be enforced? How will GM crops be kept separate? How much will this cost?

There does seem to be a reasonable way out of this conundrum. Label foods not according to the process by which they were produced, but according to the contents of the final product. If a genetically modified food is nutritionally or compositionally different from its traditional counterpart, it should be labeled as such. If there is no difference, then what is the purpose of labeling? This actually is the current point of view of the Canadian government and its scientific advisors.

So how do we respond to those consumers who say they have the right to know what they eat, even if there are no safety concerns? Fine, but why focus only on GM foods, then? What about asking for declarations about the number of insect parts or rat droppings allowed per serving (there are regulations about these), or the specific pesticides or fertilizers used, or toxins introduced by traditional crossbreeding, or whether the food was grown hydroponically? What about labeling lima beans as a source of natural cyanide? Why not put a warning on alfalfa sprouts about the risk of E. coli 0157:H7 poisoning? Shouldn't organic foods produced from crops sprayed with Bt bacteria be labeled? These bacteria release the same toxin as crops that have the Bt gene inserted. Don't consumers have the right to be informed about these things? Obviously, the labeling issue is not a simple one, and there are diverse views — though not all are of equal validity.

Dr. Andrew Weil, whose views on "natural healing" have turned him into a veritable industry, suggests "not to buy products whose labels list more chemicals than recognizable ingredients." I wonder what he thinks "recognizable" ingredients are made of?

SCIENTISTS SMELL A RAT IN
FRENCH GMO RAT STUDY

A French study published in 2012 that purports to show a link between the consumption of genetically modified corn and a variety of ailments, including cancer, was just the tasty morsel that critics of genetically modified foods (GMOs) hungered for. For many scientists, however, the study proved to be a source of indigestion.

Although California's Proposition 37, which would have required the labeling of foods that have any component derived from genetically modified crops, was defeated back in 2012, GMOs are still a hot-button issue. Emotions have boiled over with members of activist groups, such as the ridiculously named Genetic Crimes Unit, screaming about genetic crimes against humanity as they don hazmat suits to block shipments of Monsanto's transgenic seeds. They are also fond of displaying a giant "fish-corn," implying that biotechnology companies are engaged in melding fish genes with corn genes. Absurd.

Mike Adams, the self-appointed "Health Ranger" who routinely floods the Internet with stupefying diatribes on his *Natural News* website, goes even further. "I predict, but do not condone," he says, "scientists who conduct research for Monsanto being threatened, intimidated, and even physically attacked . . . an inevitable reaction to the unfathomable evil being committed by the GMO industry and its co-conspirators." Seems to me that Adams is the evil one by implanting such ideas. There are indeed some very legitimate issues to be addressed about genetic modification, but proper intellectual discourse leaves no room for such inflammatory tirades.

Mistrust and confusion are often the result of a lack of under-standing of the science involved. So let's take a look at what the controversy, at least as it pertains to the French study, is all about. The researchers aimed to explore the effects of consuming corn that is genetically modified to resist Roundup, Monsanto's popular herbicide. Such Roundup-resistant corn is unharmed when sprayed with glyphosate, the active ingredient in Roundup, while weeds wilt. This is of great advantage to growers because the technology makes weed control easier and more effective, and fields require less tillage while yields and profits increase. Before the introduction of glyphosate-resistant crops, it was common to use as many as ten different herbicides, most of which had worse toxicological profiles than glyphosate.

Glyphosate was discovered by John Franz back in 1970, while he was working at Monsanto. It works by inhibiting the plant enzyme EPSP (5-enolpyruvylshikimate-3-phosphate synthase, if you must know) which is critical for the synthesis of three essential amino acids: tryptophan, tyrosine, and phenylalanine. These in turn are needed by the plant for protein synthesis as well as for conversion into a variety of compounds such as phenolics, tannins, and lignins that are essential for plant life. If EPSP is inactivated, the plant withers and dies.

Some microbes also rely on EPSPs for protein synthesis, and in 1983, researchers discovered that a strain of the common soil bacterium *Agrobacterium tumefaciens* is highly tolerant to glyphosate because its EPSP is less sensitive to inhibition by this herbicide than the version found in plants. By 1986, the bacterial gene that codes for this enzyme was isolated and soon inserted into the genome of soybeans, corn, canola, alfalfa, and sugar beets, allowing fields to be sprayed with Roundup for elimination of weeds without affecting the crops. As a result, genetic modification has become the most rapidly adopted technology in the history of agriculture. But it has also unleashed a cavalcade of criticism.

There are concerns about seed companies establishing strict criteria for the use of their seeds by farmers, there are questions about weeds developing resistance, and, of course, worries about safety. While the majority of scientists familiar with the technology were satisfied that the concerns had been properly addressed, there were some who thought that regulatory agencies had jumped the gun. One of these was Gilles-Eric Séralini, lead author of a controversial 2012 study.

Séralini has written several anti-GMO books and has published other papers that claim to show adverse effects attributed to GMOs. He is a vocal anti-GMO activist and has already been chastised by the European Food Safety Authority (EFSA) for improper analysis of data. His 2012 study involved feeding various combinations of genetically modified corn and glyphosate to rats over their lifetime and concluded that the experimental rats had a shorter life expectancy, developed more tumors, and had more liver and kidney problems than the control group. There were horrific pictures of rats with giant tumors that were quickly snapped up by a media not adverse to sensationalism.

The response from the scientific community was immediate and harsh. The control group was way too small, there was no disclosure of control rats with tumors, data were improperly interpreted, there was no dose-response relationship, and the strain of rat used was genetically susceptible to tumors. Particularly bothersome was the fact that the research group refused to provide advance copies of their work to reporters unless they signed agreements not to consult other experts. This flies in the face of proper scientific practice. Furthermore, Séralini has now stated that he will not allow scientists from EFSA to verify his results because they are the ones who approved GMOs in the first place and therefore cannot be trusted. In light of the controversy Seralini was asked to withdraw the paper, which he refused to do. *Food and Chemical Toxicology*, the journal in which the paper appeared, then retracted it.

In any case, this study has virtually no relevance to people because the diet the rats were fed is not even remotely reflective of the human consumption of foods that have components derived from genetically modified corn. The media randomly bandies about the statement that most of the food we eat contains genetically modified ingredients. Technically, that is true if you consider, for example, high-fructose corn syrup (HFCS) derived from GM corn to be a genetically modified ingredient. The fact is that there is no vestige of genetic modification in this product. It is indistinguishable from any other HFCS. Contrary to popular belief, there are no genetically modified strawberries, tomatoes, potatoes, wheat, rice, or fruits on the market, with the exception of Hawaiian papaya, which has been engineered to protect it against a fungus, thereby saving a whole industry.

Although GM sweet corn is grown in a few places, by far the majority of GM corn goes into animal feed. Our consumption of GM ingredients is limited to some food additives and oils that are derived from GM corn, soy, or canola. This has little relation to feeding GM corn to rats as the major component of their diet. Furthermore, millions and millions of cattle and poultry have now been raised on GM corn over many generations without any health effects being noted in them or their consumers.

What we need in the GMO controversy is reasoned argument, not scandalous headlines. "Study: GMOs May Shorten Your Life" shrieks a report on Séralini's paper by Rodale Press. The study shows nothing of the kind. What it does show is the readiness of some GMO opponents to jump on a questionable study to promote their fearmongering agenda.

FROM TWITCHING WORMS TO
NON-BROWNING APPLES

The tiny worm's twitch was hardly noticeable, but with that slight shudder, science took a giant leap! A leap big enough to lead to a Nobel Prize that would pave the way to apples that will not brown, onions that will not make you cry, cottonseeds that you can eat, and diseases that you can treat.

The 2006 Nobel Prize in Physiology and Medicine was awarded to Professors Andrew Fire of Stanford University and Craig Mello of the University of Massachusetts for their discovery of RNA interference and its role in gene silencing. Genes are those segments of the "master molecule of life," DNA, that speak, but not with words. Their language is expressed in molecules, specifically ones known as messenger RNA or mRNA. The message they carry is the set of instructions for the construction of proteins.

Life is all about proteins. Not only are these molecules the building blocks of our tissues, they make up the antibodies that protect us from disease, the receptors that allow cells to communicate with each other and the enzymes that catalyze virtually every reaction that goes on in our bodies. But how do cells know which proteins to make? That's where the 30,000 or so genes dispersed along the strands of DNA come in. Each gene holds the instructions for making a particular protein, but the problem is that proteins are synthesized not in the nucleus but in the cytoplasm of a cell. How, then, does the message get from the DNA in the nucleus to the protein-making machinery in the cytoplasm? By means of the appropriately named messenger RNA. If this process is interfered with, the protein the gene codes for doesn't get made, and the gene is effectively silenced.

Now back to our little nematode worms. Some of these

creatures make twitching movements because they lack a protein needed for proper muscle function as a result of having a non-functional gene. Fire and Mello's breakthrough discovery involved making normal worms twitch by silencing the appropriate gene through injection of a special type of RNA (double-stranded RNA). It turns out that if this tailor-made RNA matches the genetic code of a specific messenger RNA, it will inactivate it, thereby essentially silencing the gene that triggered the formation of that particular messenger RNA.

Subsequent research showed that this RNA interference (RNAi) machinery can be activated in yet another fashion, without the introduction of any double-stranded RNA from the outside. Sometimes, for proper functioning of our bodies, the synthesis of certain proteins needs to be suppressed; some genes have to be silenced. Cells accomplish this through making double-stranded RNA via an intermediary known as microRNA, which in turn is synthesized on instructions encoded in the cells' DNA. In other words, DNA contains genes that can silence other genes through RNAi.

With the difficult theoretical stuff out of the way, let's get down to some practicalities. The world has no need to remedy muscular problems in worms, but how about producing apples that do not turn brown? At first this may seem like a frivolous application of RNA interference, but that is not necessarily the case. A Canadian biotechnology company, Okanagan Specialty Fruits (OSF), has developed a non-browning apple by silencing a gene that codes for an enzyme known as polyphenol oxidase (PPO).

When an apple's cells are ruptured by bruising, slicing, or biting, PPO and oxygen from the air combine with naturally occurring phenols in the apple to trigger a chemical reaction that forms melanin, a brown substance that is thought to protect the apple from attack by microbes. But the brown discoloration is unappetizing and often results in apples being discarded. The traditional way of preventing such browning is with lemon or pineapple juice, the acidity of which inactivates polyphenol oxidase. Commercially packaged apple slices are usually dipped in an antioxidant solution of calcium ascorbate. Genetically modified apples that do not brown would not require either treatment. And sliced apples that do not brown would avoid the yuck factor and make for a healthy addition to children's lunches. Any method that allows for greater apple consumption is attractive.

The exact fashion in which the Arctic Apple, as it will be known, is genetically modified is proprietary information, but it is accomplished through RNA interference. Here is a possible way. Some apples are naturally very low in polyphenol oxidase because they express a gene that codes for the double-stranded RNA that in turn silences the PPO gene. Through standard genetic modification methods, this silencing gene can be copied and inserted into the DNA of other apples with the result that PPO production will be silenced and the apples will not turn brown.

Not everyone is thrilled by the possibility of genetically altering apples in this fashion. Organic growers worry that pollen from the modified apple trees will spread to their orchard, potentially causing them to lose their organic status. Okanagan Specialty Fruits argues that apple pollen does not blow around easily, and the chance of it spreading to a neighboring orchard is slim. Some critics, particularly anti-GMO activists, have suggested that silencing the PPO gene may have unintended negative consequences, but there is no evidence for this. That comes as no surprise because there are no novel proteins being formed. Field trials have shown that the modified apples are like all other apples except that they do not turn brown.

Using RNA interference technology, the lachrymatory factor synthase gene in onions can be silenced so that the nutritional qualities of this vegetable can be enjoyed without weeping.

And how about cottonseed? The world produces some 44 million tons of high-protein seed every year that cannot be eaten because it contains the poisonous compound gossypol. Using RNA interference, the gossypol-producing gene can be silenced and enough protein to meet the daily requirements of half a billion people can be produced. But perhaps the most alluring potential of RNA interference lies in tackling genetic diseases. There have already been some preliminary successes, albeit only in mice, with silencing genes that code for toxic proteins such as the ones found to be present in Huntington's disease, as well as in silencing genes that cause high cholesterol levels.

Regulatory agencies in Canada and the U.S. have ruled that the Arctic Apple is safe and can be marketed although so far it has only appeared as packaged apple slices. There is no doubt that the journey from twitching worms to non-browning apples has been a fascinating one! Let's hope we won't end up with worms in the apple by silencing the polyphenol oxidase gene.

THE SAGA OF GOLDEN RICE

You really will see better if you eat carrots. But there's a catch. Carrot therapy only works if your vision problems are due to a deficiency in vitamin A. This is a rarity in North America, but sadly not in the developing world. An estimated 250,000 to 500,000 cases of childhood blindness are caused by a diet that is deficient in vitamin A, or in its precursor, beta-carotene, which the body can convert to the vitamin.

What does vitamin A have to do with vision? Retinol, as the vitamin is also known, is absorbed from the digestive tract and is chemically modified to become retinal in the body. The retinal then complexes with a protein in the eye that is known as opsin. When light hits the opsin-retinal complex, a chemical change ensues, unleashing a cascade of events that lead to the transmission of an impulse up the optic nerve. Given that vitamin A is found in meat and fish, a dietary deficiency in North America is unlikely. Even vegetarians are safe. Although vitamin A occurs only in animal products, our bodies can make it from beta-carotene, the orange-colored molecule found in carrots and numerous other vegetables. Rice, however, has virtually no beta-carotene, and as a consequence, vitamin A deficiency in rice-based societies, such as India, China, and Indonesia, is common. As a result, these populations experience widespread childhood blindness, and tragically, more than half of those who lose their sight die within a year.

Various attempts have been made to supplement the diet with vitamin A. In Indonesia, for example, it has even been added to packets of the widely used flavor enhancer MSG. But the problem persists. That's why so much excitement was generated in 2000, when recombinant DNA technology made possible the insertion into rice of a gene that codes for the production of beta-carotene. This gene, taken from daffodils, allowed the newfangled rice to produce enough beta-carotene to actually color it yellow — hence the term "golden rice."

Proponents of genetically modified crops highlighted the development of golden rice as a breakthrough and suggested it would be a useful way to put a dent in the vitamin A deficiency problem. Opponents pointed out that the amount of beta-carotene — roughly one and a half micrograms per gram of rice — was too little to have any practical impact. They claimed that the whole golden rice issue was an industry ploy to push for wider acceptance of genetic modification. Researchers countered that the technology was new and that improvements would surely be forthcoming. And they were right! A team at Syngenta Seeds in

Britain found that a gene taken from corn and inserted into rice is far more adept at churning out beta-carotene than the original one from daffodils. This second-generation golden rice contains almost twenty-five times as much beta-carotene as the original version. A typical daily serving of 200 grams could then provide the minimal vitamin A requirement. If only countries where vitamin A deficiency is common made it available!

Rice enhanced with beta-carotene can do more than help with visual problems. Vitamin A deficiency can lead to abnormal bone development as well as a greater susceptibility to infections. Low blood levels of vitamin A have even been linked with an increased risk of cancer. And should you think that Syngenta is just an evil multinational, trying to capture the rice market in developing countries with overly optimistic promises, know this: the company has donated the rights for golden rice to the non-profit Humanitarian Rice Board, which will make it available to farmers for free. India and the Philippines have already approved trial plantings, despite objections from anti–genetic modification groups that claim golden rice is a pie-in-the-sky approach and will not solve the vitamin A deficiency problem. But scientists have never claimed it would. Golden rice is just one method of providing extra vitamin A. And we do have to be impressed by the fact that, in just a few years, researchers found a way to increase the beta-carotene content of golden rice twenty-five-fold!

These potential advances are not limited to rice. In India, there is hope that a genetically modified potato will help combat malnutrition. Much of the population is vegetarian, but pulses and legumes, the main sources of protein, are expensive and often in short supply. Potatoes can be grown easily, but they don't contain much protein. This can be remedied, however, through the addition of a gene isolated from a South American plant known as amaranth. The gene in question codes for the production of a protein rich in the essential amino acids lysine and methionine. Too little methionine in the diet is known to affect brain development. Amaranth is commonly eaten in South America, so the transfer of a gene from it into potatoes does not present a health risk.

Admittedly, many people remain suspicious of GMO technology in spite of its potential to address nutritional problems. They don't want their food genetically modified. Little do they realize that virtually everything we eat has been modified, although not necessarily through the use of recombinant DNA. Centuries of crossbreeding, as well as treatment of seeds with chemicals or radiation to induce mutations, have resulted in extensive genetic modification of plants. In most cases, thousands of genes with unknown function may be involved. People don't worry about this (and shouldn't), yet they become extremely concerned when a single specific gene with a known function is transferred. I guess this isn't too surprising, given that surveys show that 43 percent of Americans believe that only genetically modified tomatoes contain genes. Maybe if they ate more carrots, modified to contain vitamin A, they would see this situation better.

AN ANTIDOTE TO THE POISONOUS TOMATO LEGEND

I ate my first tomato when I was about twelve years old. Actually, it wasn't even a tomato; it was tomato sauce on a pizza. I really don't know why, but growing up, I had a real aversion to tomatoes as well as to any food that contained them. And then a friend convinced me to try pizza. All of a sudden, a whole world opened up! Tomatoes, I discovered, taste great! And, as I would eventually learn, they were pretty healthy to boot. Perhaps this is why now, in my public lectures on food, I like to tell the tale of Robert Gibbon Johnson's tomato escapade.

As the story goes, in Salem, New Jersey, back in 1820, Colonel Robert Gibbon Johnson took out an ad in local newspapers inviting the public to gather in front of the courthouse on a Sunday afternoon to experience an epic event. He promised that, in full view of all, he would eat a tomato! At the time, tomatoes were thought to be poisonous, and people were drawn by the possibility of seeing someone do harm to himself. Johnson, though, knew better, and thought Americans were depriving themselves of a delicious fruit. (Yes, the tomato, being the seed-bearing part of a plant, is a fruit.)

Colonel Johnson hired a little band to play a funeral dirge in the background as he picked up a tomato and took a large bite. Perhaps to the disappointment of the crowd, he didn't clutch his chest, foam at the mouth, or drop to the ground. He survived, and on that day the tomato industry was born, and we are all better for it. At least, so goes the oft-repeated story. Alas, repetition does not make a story true, no matter how compelling it may be.

Johnson really did exist. He was an elected member of the New Jersey State Assembly and at one time served as president of the Salem Horticultural Society. But in none of his writings did he ever make any mention of the tomato-eating incident. Indeed, we don't hear of the supposed epic moment until 1937, when it is highlighted in Joseph Sickler's book *The History of Salem County*. It seems Salem's history needed a bit of spicing up and the "toxic tomato" story happened to fit the bill. But the truth is that by 1820, tomatoes, which are actually native to America, were regularly eaten. Long before Johnson's mythical, foolhardy experiment, President Thomas Jefferson had grown tomatoes in his own garden.

Myths are often born out of smidgens of facts. Sickler's poisonous tomato account may have been triggered by the close similarity of the tomato plant to others of the nightshade family, such as belladonna, which truly are poisonous. Nightshade plants contain potential poisons such as atropine in the case of belladonna, solanine in green potatoes, and tomatine in tomatoes. But of course, in toxicology, dosage is critical. Given that in a large enough dose, it can cause harm, technically tomatine can be considered a poison, but the amount present in a tomato is negligible. In any case, the compound is found mostly in the flowers and leaves of the plant, which are not eaten. Tomatoes are not toxic, and there is no evidence that anyone ever thought they were.

There are current myths about tomatoes as well. A popular one contends that some have been genetically engineered to keep them from freezing by inserting a gene from a fish known as the Arctic flounder. There are cartoons galore depicting tomatoes with fins or fish that have tomatoes for heads. While a genetically engineered tomato known as the Flavr Savr was briefly marketed with claims of improved taste (which it did not have), there are now no genetically engineered tomatoes on the market. Some researchers have indeed played with the possibility of inserting a fish gene that codes for an "antifreeze" protein into tomato plants to keep the fruit from freezing in case of a sudden cold spell, but this research has not borne fruit. In any case, this would not make the tomato into a fish-fruit hybrid. Fish have over 30,000 genes, and no one gene makes a fish a fish. A tomato with one fish gene would still be a tomato. There could, however, be an issue with allergies, and if this project were ever to be commercialized, extensive testing would have to be done to ensure that someone with a fish allergy would not react to the engineered tomato.

A more realistic controversy about tomatoes focuses on whether organically grown tomatoes are in any way superior to conventionally grown ones. Like other fruits and vegetables, tomatoes contain compounds with antioxidant properties. While popular books and magazines loudly tout the almost-magical health benefits of antioxidants, the scientific literature is less compelling. There is no doubt that eating large amounts of fruits and vegetables is beneficial, but the exact reason is not clear. Produce contains thousands of compounds with biological properties, and specifically which of these compounds, if any, are responsible for the benefits is not known. Polyphenols, though, are reasonable candidates.

A Spanish study has shown that organic tomatoes have a slightly higher polyphenol content than those grown by conventional means, which is not surprising. Plants do not produce polyphenols for the benefit of humans. They produce the chemicals to help them survive stressful conditions such as a lack of nutrients in the soil or attack by insects and fungi. Without the application of synthetic pesticides and fertilizers, tomato plants are more stressed and produce more polyphenols. But this is unlikely to have any practical implications. In the context of an overall diet, the small differences between the polyphenol content of organic and conventionally grown produce is irrelevant. Implying that organic tomatoes are better for us because they may have a slightly higher polyphenol content, especially in the face of a lack of studies showing that increased polyphenol content leads to better health, has about as much merit as the story about Robert Gibbon Johnson surviving the consumption of supposedly toxic tomatoes.

ORGANIC AGRICULTURE

There were piles of all sorts of tomatoes in the produce aisle of the supermarket. The ones that caught my attention sat neatly wrapped in plastic in groups of four. They weren't any better looking than the others, but they were certainly more expensive! What sort of tomatoes were these, to command a king's ransom? Well, they were "organic." Why did they warrant the investment? Because, as the label declared, "when you purchase organic produce, you are taking part in the healing of our land, the purifying of rivers, lakes, and streams, and the protection of all forms of life from exposure to chemicals used in conventional farming." Surely only a callous chemist with a disregard for nature would purchase any other sort of tomato.

There is no doubt that the organic produce market is growing. Some buy organic because they believe such foods are healthier; others do so to help save the environment from those nasty agrochemicals. These beliefs are certainly worth investigating. But what does "organic" actually mean? Essentially, organic food must be produced without the use of synthetic pesticides, artificial fertilizers, antibiotics, or growth-promoting hormones. Genetically modified organisms are not allowed, and irradiation cannot be used to control bacteria. Sounds just like farming roughly 100 years ago. Back then, feeding the masses required the involvement of some 70 percent of the population in farming in some way. Yields were low, crop losses to insects, fungi, and weeds were high. That's why farmers welcomed the introduction of scientifically designed fertilizers and pesticides. That's why, today, 2 percent of the population can feed the other 98 percent.

Such advances have not come without a cost. Pesticides and nitrates from fertilizer enter ground water, posing potential environmental and health consequences. So people hark back to the "good old days," when food was untainted and people lived in blissful health. Of course, those "good old days" only exist in people's romanticized imaginations. Food borne diseases were rampant, and fresh fruits and vegetables in winter were virtually unheard of. Nutrient deficiency diseases cut a wide swath through the population. Of course, not even the greatest advocates of organic agriculture suggest that we can realistically turn back the clock and provide food for the world's population using only organic methods. They claim a niche market that caters to people who are conscious of their environment and health.

So, do consumers who buy organic avoid pesticides? Hardly. Organic farmers are allowed to use a number of pesticides, as long as they come from a natural source. Pyrethrum, an extract of chrysanthemum flowers, has long been used to control insects. The Environmental Protection Agency in the U.S. classifies it as a likely human carcinogen. There you go, then, a carcinogen used on organic produce! Does it matter? Of course not. Just because huge doses of a chemical, be it natural or synthetic, cause cancer in test animals does not mean that trace amounts in humans do the same. Furthermore, pyrethrum biodegrades quickly, and residues are trivial. But that is the case for most modern synthetic pesticides as well.

Organic farmers are also free to spray their crops with spores of the *Bacillus thuringiensis* (*Bt*) bacterium, which release an insecticidal protein. Yet organic agriculture opposes the use of crops that are genetically modified to produce the same protein. Isn't it curious that exposing the crop to the whole genome of the bacterium is perceived to be safe, whereas the production of one specific protein is looked at warily? The truth is that the protein is innocuous to humans, whether it comes from spores sprayed on an organic crop or from genetically modified crops. True, organic produce will have lower levels of pesticide residues, but the significance of this is highly debatable.

A far bigger concern than pesticide residues is bacterial contamination, especially by potentially lethal *E. coli* 0157:H7. The source is manure used as a fertilizer. Composted manure reduces the risk, but any time manure is used, as is common for organic produce, there is concern. That's why produce should be thoroughly washed, whether conventional or organic. Insect damage to crops not protected by pesticides often leads to an invasion by fungi. Some fungi, like *Fusarium*, produce compounds that are highly toxic. In 2004, two varieties of organic cornmeal had to be withdrawn in Britain because of unacceptable levels of fumonisin, this natural toxin.

Are organic foods more nutritious? Maybe, but marginally. When they are not protected by pesticides, crops produce their own chemical weapons. Among these are various flavonoids, antioxidants that may contribute to human health. Organic pears and peaches are richer in these compounds, and organic tomatoes have more vitamin C and lycopene. But again, this has little practical relevance. When subjects consumed organic tomato purée every day for three weeks, their plasma levels of lycopene and vitamin C were no different from those seen in subjects who consumed conventional purée. Where organic agriculture comes to the fore is in its impact on the environment. Soil quality is better, fewer pollutants are produced, and less energy is consumed. But we are simply not going to feed 7 billion people organically.

Finally, do organic tomatoes taste better? I can't tell you. Instead of shelling out $5.80 for four tomatoes, I bought a bunch of regular tomatoes, some apples, and some oranges for the same total. And I think I got a lot more flavonoids and vitamins for my money.

PESTICIDE PROBLEMS

Pesticides are designed to kill. Of course, they are designed to kill the insects, the fungi, the rodents, and the weeds that compete for our food supply, that carry disease, or that tarnish our green space. But they can also kill people. And, unfortunately, that isn't a rare occurrence. The World Health Organization estimates that there are roughly three million cases of pesticide poisoning worldwide every year, and close to a quarter-million deaths! Astoundingly, in some parts of the developing world, pesticide poisoning causes more deaths than infectious disease. How? Certainly, people do die from a lack of proper protective equipment, or because they can't read the instructions about diluting the chemicals properly. But the real tragedy is that the main cause of death due to pesticides is suicide.

Believe it or not, about a million people in the world do away with themselves every year. More than three-quarters of these are in developing nations, where life can be so difficult that the alternative may seem more attractive. In Sri Lanka, suicide is one of the top causes of death among young people, and in China, more young women kill themselves than die from other causes. Pesticides are the weapons of choice. In rural Sri Lanka, pesticide poisoning is the main cause of death reported in hospitals. There are wards devoted to patients who have tried to kill themselves with organophosphates, one of the most toxic classes of pesticides. In 1974, when paraquat was introduced in Samoa, suicide rates went up, sharply. They dropped back down in 1982, when paraquat was taken off the market. In Amman, Jordan, poisonings fell way off when parathion was banned. Obviously, if the use of the most toxic pesticides could be curtailed in these countries, many lives would be saved. Sadly, though, these chemicals are often completely unregulated. Some of the most toxic ones are readily available in stores and will be sold to the illiterate farmer who has virtually no chance of using them properly. Pesticide companies, in some cases, pay their salespeople on commission, so it is in their interest to push product even when it may not be necessary. In Sri Lanka, pesticides are advertised on radio to the public, often painting an unrealistic picture of magical, risk-free crop protection. Some sort of joint effort by pesticide manufacturers and governments is needed to keep the most toxic pesticides out of developing countries.

In North America, our pesticide regulations are far more stringent, and farmers must be licensed to use these chemicals. That doesn't mean we don't have problems. In North Carolina, for example, thousands of migrant workers are employed on tobacco, vegetable, fruit, and Christmas-tree farms. Many of them live in dilapidated housing next to the agricultural fields, and their homes and bodies are contaminated with pesticides. Metabolites of organophosphates commonly show up in their urine. This is not surprising, given that access to showers and clean clothes after working in the fields is limited. Even though there may be no immediate effects of such exposure, sufficient studies have suggested ominous links — between pesticide use and neurological problems, developmental delays, Parkinson's disease, and cancer — to cause concern. What's the answer? Elimination of agricultural pesticides is simply not an option. But providing workers with safe housing, clean clothes, showers, and above all, pesticide safety training certainly is.

Of course, working in the fields of North Carolina is not the only way to be exposed to pesticides. Garden-supply stores sell a wide array of such products. They are all "registered," meaning that they have undergone extensive safety evaluation. Risks should therefore be minimal, if the products are properly used. That, though, is a big "if." An often-quoted study at Stanford University found a link between Parkinson's disease and domestic pesticide use. People with as few as thirty days of exposure to home insecticides were at significantly greater risk; garden insecticides were somewhat less risky. Because of the large variety of products available, the researchers were not able to zero in on any specific ingredients. Another study, this one at the University of California at Berkeley, compared pesticide exposures of children diagnosed with leukemia to a healthy control group matched for age and socioeconomic status. The families of children with leukemia were three times more likely to have used a professional exterminator. During pregnancy, exposure to any type of pesticide in the home coincided with twice as much risk. But — and an important "but" — there was no association between leukemia and pesticides used outside the house! Yet I have often seen activists who oppose cosmetic lawncare chemicals use the leukemia argument to demonize this practice.

Pesticides cannot all be lumped together, in terms of their safety profile. There are tremendous differences between the various insecticides, which differ extensively from herbicides and fungicides. And one must always remember that associations cannot prove cause and effect. Physicians should realize this, one would think. Apparently, not all do. In a letter to a medical publication, a doctor chastised the federal government for allowing people to be exposed to dangerous substances on their lawns, and buttressed the argument with this example: "A boy was removed from a daycare three years ago because his parents noticed the lawn was being treated with pesticides and the child began to suffer health problems and recurrent pneumonias. He developed acute lymphoblastic leukemia." The simple-minded message, of course, is that the spraying caused the leukemia — a gigantic, and inappropriate, leap of faith.

Great caution must be used with insecticides in the home, and I believe that their use during pregnancy should be totally avoided. But using insecticides inside a house presents a completely different scenario from occasionally spraying a lawn with fertilizer and weed killer. Different chemicals, different exposures, different risks. When contemplating the use of pesticides, always remember that, while there may be no completely safe substances, there are ways to use substances safely.

THE EVOLUTION OF HERBICIDES

Charles Darwin is of course best known for his epic work *On the Origin of Species*, but in 1881, he also carried out a landmark study that would lead to the discovery of plant hormones. Although it had long been known that plants tend to bend towards a light source, nobody had explored how this was actually happening. Darwin decided to find out. Experimenting with grass seedlings, he covered different parts of the growing leaf with an opaque material and found that if the tip was covered, or totally removed, there was no tendency to bend towards the light. The conclusion was that the tip was responsible for sensing light and that it then sent some sort of message down to the lower part of the leaf to trigger bending.

Just how that message was transmitted was eventually discovered in 1928 by Dutch botanist Frits Went. He found that light triggered the formation of chemicals in the tip that then scooted through the plant causing structural changes that resulted in bending towards the light. These chemicals were christened "auxins" from *auxein*, the Greek for "grow." Within a few years, Harvard botanist Kenneth Thimann managed to isolate a major auxin, indole-3-acetic acid, that was responsible for plant growth.

The existence of auxins germinated the idea that auxin-like compounds could serve as herbicides by stimulating such rapid growth that the plant would outgrow its nutrient supply and die. This indeed turned out to be a valid theory and resulted in the formulation of the herbicides 2,4-D and 2,4,5-T, compounds that became infamous for their inclusion in Agent Orange, the notorious defoliant used during the Vietnam War. Although not known at the time, 2,4,5-T was contaminated with a side product of its synthesis, 2,3,7,8-tetrachlorodibenzodioxin, or TCDD. This was later found to be not only toxic but carcinogenic, and its production was phased out in the 1970s. On the other hand, 2,4-D does not raise the same concern since its synthesis does not lead to any TCDD as a side product. It has proven to be a highly effective commercial weed killer, best known as Killex.

Another auxin analogue, 3,6-dichloro-2-methoxybenzoic acid, was introduced as a herbicide in 1967 under the name dicamba. Like 2,4-D, it is effective against broadleaf plants but not narrowleaf ones like grasses. This means that it can be used on lawns and golf courses as well as on cereal crops such as wheat, barley, rye, and oats. Dicamba cannot be used to control weeds in soybean or cotton fields because these are broadleaf plants that are susceptible to the herbicide. However, genetic engineering has found a solution to this problem.

A bacterium discovered in soil, *Stenotrophomonas maltophilia*, produces enzymes that can decompose dicamba. The gene responsible for one of these enzymes has been successfully incorporated into the genome of soybean and cotton seeds, meaning that plants grown from these seeds will be resistant to dicamba. Fields can then be sprayed with this herbicide to kill weeds without harming crops. Obviously, this technology has been welcomed by many farmers, but not all. There is no question of toxicity because dicamba is remarkably nontoxic to humans.

But there is a problem of the spray drifting and affecting nearby fields where crops that are not dicamba-resistant are grown.

Organic farmers would obviously be concerned since organic agriculture does not allow for the use of genetically engineered seeds. In this case any drift of dicamba could result in major crop losses. And this has happened, resulting in a number of lawsuits. Some farmers have claimed that they have had to switch to dicamba-resistant seeds against their will in order to have a viable crop. They then have to inform their neighbors of this, who then also are put into a situation where they have to switch to the genetically engineered seeds. Lawsuits have been filed with the claim that manufacturers of these seeds, such as Monsanto and BASF, have violated antitrust laws by driving competitors out of the market. That's because farmers were worried about the risk of drifting dicamba and would not purchase seeds that were not dicamba-resistant.

Monsanto argues that it has developed formulations that have greatly reduced volatility, preventing drift, although some scientists contest this. Governments have also introduced stringent regulations on when and how dicamba can be sprayed. Whether these regulations are obeyed is questionable because they are difficult to follow. There are other concerns as well. Weeds are already showing signs of developing resistance to dicamba, and it seems that dicamba drift may even be affecting oak trees.

There is no question that the use of dicamba is a hot topic. More than hot. Life-threatening. In 2016, a farmer who lost much of his soybean crop attributed the loss to spray drifting from a neighboring farm and confronted the neighbor, physically attacking him. The victim then pulled a gun, and in an act of what he claimed was self-protection, shot and killed the aggressor. He was convicted of second degree murder.

SCIENCE BY PETITION

I'm sure Lisa Leake is a well-meaning young lady and a fine mother. But I think she could use a lesson in chemistry. Lisa and fellow food blogger Vani Hari were the movers and shakers behind a petition to "remove all dangerous artificial food dyes" from Kraft's classic macaroni and cheese and replace them with the natural dyes used in the United Kingdom. The argument is that Yellow #5, also known as tartrazine, and Yellow #6, known as Sunset Yellow, may be linked to health problems, whereas the natural colors, namely beta-carotene and paprika, have no such dark clouds hanging over their heads.

Food dyes have always been one of the most controversial classes of food additives because they serve only a cosmetic purpose. They do not contribute anything nutritionally and in fact may make foods of poor nutritional quality more appealing. Before getting back to the petition and allegations of danger, a little history seems to be in order.

As early as the first century, Pliny the Elder noted that wine was sometimes artificially colored, possibly with squid ink. Saffron, paprika, turmeric, beet extract, and various flower petals have long been used to color foods. A peasant in the Middle Ages may well have eaten bread that was adulterated with lime to mimic refined flour, preferred by the rich but unavailable to the poor. King Edward I (1239–1307) took such adulteration very seriously and issued an edict that a baker guilty of such an offense should be dragged "through the great streets where there may be the most people assembled . . . with the faulty loaf hanging around his neck." Should he repeat the offense, he would be pilloried for an hour, and if he still didn't learn his lesson, "the oven shall be pulled down and the baker made to foreswear the trade in the city forever."

In the fourteenth century the coloring of butter was made illegal in France, and a law passed in 1574 forbade the coloring of pastries to simulate the presence of eggs. Copper compounds were commonly used to "green up" vegetables. In 1820, English chemist Friedrich Accum recounted the misadventures of a "young lady who amused herself while her hair was dressing with eating sapphire pickles impregnated with copper." The episode did not have a happy outcome. "She soon complained of pain in the stomach. In nine days after eating the pickle, death relieved her of her suffering." Accum also documented the use of red and white lead, vermilion (a mercury compound), and copper arsenite in candies designed to appeal to children. He actually published the names of the guilty manufacturers, making some powerful enemies in the process, yet adulteration continued unabated.

The situation was certainly no better in America. Pickles were bathed in copper sulfate and milk was tinged yellow with lead chromate. Indeed, this was such a common process that when white milk was available, people refused to drink it, thinking it had been adulterated. But even back then, there were consumer advocates. At the 1904 Saint Louis Exposition, they displayed silks colored with dyes used by food manufacturers, implying that chemicals that could be used to dye fabrics were not suitable for consumption. This is what I would refer to as a "fallacy by association." Palm oil, for example, is used to make napalm, but that has nothing to do with its safety as a food. Similarly, food dyes cannot be declared dangerous just because they are made from petroleum, a substance no one would ever want to consume.

While I have no problem urging a reduction in the use of food dyes, I have a problem with unscientific arguments used towards this end, such as the petitioners' claim that "food companies [are] feeding us petroleum disguised as brightly colored food dyes." To dig an even deeper hole, the bloggers list seven reasons why they hate food dyes, with number one being that "they are made in a lab with chemicals derived from petroleum, a crude oil product, which also happens to be used in gasoline, diesel fuel, asphalt, and tar." This is senseless fear mongering from a young mom lacking any scientific background.

The allegation that we are being fed petroleum disguised as food dye is blatantly absurd. Food dyes, while synthesized from compounds found in petroleum, are dramatically different in molecular structure from any petroleum component. Furthermore, the safety of a chemical does not depend on its ancestry but on its molecular structure. And the way to evaluate safety is through proper laboratory and animal studies with continued monitoring of human epidemiology.

Over the years, as testing methods became more and more sophisticated, and regulations more stringent, many of the dyes used historically by the food industry were removed from the market. The ones that remained, such as tartrazine and Sunset Yellow, have passed the scrutiny of regulatory agencies and are allowed in a wide variety of foods.

The legal use of a food dye, however, cannot guarantee that no adverse effect will be noted. There is always the possibility that a small subset of the population will experience some adverse reaction. Allergic reactions as well as behavioral problems in children have been noted with some food dyes, although there is a divergence of opinion about the seriousness of the problem. It is such uncertainty that has prompted the petition against Kraft. While the goal is reasonable, the suggestion that food dyes are a problem because they are man-made chemicals derived from petroleum is not.

Though it is true the dyes used by Kraft in North America have all passed through the regulatory hoops and hurdles, there have been enough questions raised about them to give us reason to evoke the precautionary principle, which states that even if there is no proof of harm, a chemical with some potential for harm should be replaced if a safer alternative is available. Paprika and beta-carotene are indeed better choices. Of course that still begs the question of why macaroni and cheese should be colored in the first place. Anyway, all this has put me in a mood for some mac and cheese. Made from scratch. No color needed. Yum!

A GRAIN OF SALT

Reducing sodium intake has been a nutritional mantra for decades. We have repeatedly been told that cutting back on salt lowers blood pressure, which in turn lowers the risk of heart attacks and strokes. But these days it seems to be in vogue to question almost every type of dietary advice that has been dispensed by health authorities, including salt intake. Questioning current dogma of course isn't a bad thing, after all, that is how science progresses. The truth is that often the evidence for recommendations is not as robust as it is made out to be, and we have seen views change about the likes of saturated fats, eggs, and sugar in our diet as new data emerge. Today, with studies being cranked out at a frantic pace it is possible to find "evidence" for almost any view that one holds, but conclusive evidence, particularly when it comes to diets, is elusive. When it comes to food, the gold standard, the randomized double-blind trial, is extremely difficult to design and carry out.

In the case of sodium, a meaningful trial would mean following groups of subjects for many years and noting the incidence of cardiovascular disease, with the only difference between groups being the amount of sodium in the diet. It is difficult enough to do this over the short-term, but that actually has been done. The famous Dietary Approaches to Fight Hypertension (DASH) trial managed to test three different levels of sodium intake by providing subjects with all their meals. They consumed either 1,500, 2,300, or 3,500 milligrams of sodium a day, with results showing a clear link between blood pressure and sodium intake. The 3,500 milligrams level was chosen because it represents the amount of sodium that is consumed on the average by the population. This translates to about 9 grams of salt (sodium chloride), or one and a half teaspoons, most of which comes from processed foods.

The trial lasted only sixteen weeks, too short to note a difference in disease patterns. As critics pointed out, demonstrating a decrease in blood pressure with reduced sodium is not the same as showing a decrease in the risk of a heart attack or stroke. But given that there is overwhelming evidence from population studies that high blood pressure is associated with cardiovascular disease, it is reasonable to recommend a cutback on salt. The question is by how much?

That question arises because some recent studies have suggested an increased risk of adverse health outcomes associated with sodium intake even in the lower 1,500 to 2,300 milligrams a day range. This, however, may have nothing to do with sodium. It is possible that people with cardiovascular disease, who have been advised to dramatically reduce their salt intake, fall into this range and suffer problems because of the preexisting condition rather than their low sodium intake. In any case, for the general population, the 2,300 milligrams target is reasonable. Debates about low sodium levels presenting a risk may have academic interest but have little practical value. The 1,500 milligrams target is unattainable for most people and given that our average intake is in the range of 3,500 milligrams a day, emphasis has to be placed on reducing this rather than worrying about too little sodium.

Cutting back isn't easy. Producers cater to our fondness for salt by adding it liberally to a wide array of foods. A bowl of cereal contains about 300 milligrams of sodium, a single hot dog can have 800, a slice of bread 230, a cup of cottage cheese 900, a couple of slices of processed cheese 700, and half a cup of commercial tomato sauce 600 milligrams. A slice of pizza can weigh in anywhere from 600 to 1,500 milligrams of sodium per slice! Obviously, it isn't hard to surpass 2,300 milligrams. So there really is no worry about consuming too little sodium; that isn't happening in the real world. There is another reason we can dismiss the naysayers who claim that the evidence to support a low sodium diet is too weak. Cutting back on sodium means a decrease in processed food intake and an increase in fruits and vegetables. And there can be no argument against that.

Another line of attack vilifies table salt for its additives and for being dried at a temperature that "radically and detrimentally alters the chemical structure of the salt." Salt is composed of sodium and chloride ions and once dissolved has no "chemical structure." As far as additives go, small amounts of ferrocyanide, phosphates, or silicates are used to ensure easy pouring, and potassium iodide is added to supply the body with the iodine the thyroid gland needs to synthesize its hormones. Indeed, in North America, goiter due to iodine deficiency has been virtually eliminated. Although potassium iodide is relatively stable, it does slowly react with oxygen to yield iodate. A small dose of sugar added to salt protects the iodide by reacting with oxygen and in a sense sacrificing itself to prevent iodide from being oxidized.

Can salt in food be "neutralized"? I mention this because I was asked about the possibility of "neutralizing" the salt on French fries by adding vinegar, the question apparently being sparked by the observation that vinegar can remove salt from winter boots. The only way to "neutralize" salt is to not add it in the first place. But remember that the salt added from a salt shaker makes up only about 14 percent of our total intake. Processed foods like salami are the real problem. Indeed, the word salami derives from the Latin meaning "salted things." And we eat way too many salted things.

As is often the case with nutritional controversies, pseudo-science slithers into the picture. In this case it is in the form of "natural" alternatives to table salt with insinuations of health benefits. Himalayan salt, which is composed of large grains of rock salt mined in Pakistan, is touted as a healthier version because it contains traces of potassium, silicon, phosphorus, vanadium, and iron. The amounts are enough to color the crystals, giving them a more "natural" appearance, but are nutritionally irrelevant. Some promoters make claims that are laughable. Himalayan salt, they say, contains stored sunlight, will remove phlegm from the lungs, clear sinus congestion, prevent varicose veins, stabilize irregular heartbeats, regulate blood pressure, and balance excess acidity in brain cells. One would have to have a deficiency in brain cells to believe such hokum. It doesn't even rise to the level of taking it with a grain of salt.

THE NO CONCLUSION
CONCLUSION

A book such as this usually ends with a conclusion. The trouble is that when it comes to the relationship between food and health, the only real conclusion is that there is no conclusion. At least not a conclusion that is truly conclusive. That's because there is always some novel nutritional claim to take into account, some new study to gauge, some new product to evaluate.

While it is true that definitive answers to questions about diet are elusive, the public perception is that there is much more confusion about nutrition than there actually is. Some of that confusion is due to the common opinion that nutrition is characterized by the "one day this, next day that" phenomenon. One day headlines stress that eggs cause heart attacks, the next day there is a story about how they may save us from having one. One day B vitamins give us energy, the next day they are associated with hip fractures. One day calcium propionate is an ideal mold inhibitor in bread, the next day it leads to weight gain, at least in mice. All of these studies may have been carried out properly in their own right, but none of them should lead to any change in dietary behavior. Recommendations should not be based on single studies; they should be based on scientific consensus that is arrived at by taking all existing information into account.

Believe it or not, scientific consensus has not changed much over the last few decades. In 1977, the U.S. introduced the first truly comprehensive dietary guidelines as put forth in the McGovern report. It recommended increased consumption of complex carbohydrates, as in whole grains, reducing added sugar to less than 10 percent of total calories, reducing overall fat, reducing saturated fat to less than 10 percent of total calories, reducing cholesterol-containing foods, reducing salt intake to five grams a day, and avoiding becoming overweight by watching calorie intake and increasing energy expenditure through exercise. Current recommendations are very much along the same lines, although more specific advice is given, such as meeting these goals by relying more on plants than animals as a protein source, consuming five to seven servings of fruits and vegetables a day, and minimizing processed foods.

Providing more dietary detail is notoriously difficult. Nutritional studies are very challenging to carry out. When it comes to diets, you can either "observe" or "perturb." An observational study is relatively easy but can only show associations. It cannot prove a cause and effect relationship. For example, one can note that populations consuming a lot of processed meats such as cold cuts or hot dogs have higher rates of colon cancer, but that does not prove that these foods cause the condition. It may be that the problem is not what the people are eating, but rather what they are not eating, that is fruits and vegetables. Or they may have different activity levels. Or they may be more inclined to smoke or drink alcohol. So more information would be needed before concluding that processed meats cause colon cancer.

That can come from case-control studies in which people who have a disease are compared with those who are healthy. Their past dietary history is investigated, and if it turns out that colon cancer victims have eaten more processed meats, then we start paying more attention to the role that such products may play in the etiology of the disease. However, to prove that processed meats cause colon cancer, we would have to enlist a number of similar subjects, divide them into two groups, and perturb the diet of one with regular doses of processed meats. These groups would then have to be followed for decades and the incidence of colon cancer statistically analyzed. Obviously, such trials present logistical and economic challenges. Furthermore, monitoring the extent to which people keep to the prescribed diet would be very difficult.

Even if these challenges were met, and even if it turns out that there are significantly more cases of colon cancer in the experimental group than in the control group, we still could not conclude with certainty that processed meats cause colon cancer. That's because it is not possible to have two diets with the only difference being the consumption of processed meats. The control group would have to have a diet in which the processed meats are replaced by something of equivalent calorie and nutrient content, which means that whatever differences are found cannot be conclusively attributed to the processed meats.

Although randomized controlled trials are difficult to carry out, they are not impossible. Researchers at the National Institutes of Health in the U.S., prompted by the increasing number of observational and epidemiological studies linking processed foods to poor health outcomes, managed to carry out such a trial. They enlisted ten male and ten female volunteers who agreed to spend a month in a metabolic ward where all their meals would be provided and their weight monitored. In a random order, for two weeks on each diet, they consumed either meals made with ultra-processed foods that contained an array of additives or meals of minimally processed foods. The key was that both diets had virtually identical calorie, sugar, fat, and total carbohydrate content. Subjects could eat as much or as little as they wanted.

A typical day on the "ultra-processed" diet started with a breakfast of Honey Nut Cheerios, whole milk with added fiber, and a blueberry muffin with margarine. Lunch was beef ravioli, parmesan cheese, white bread, margarine, diet lemonade with fiber, and oatmeal raisin cookies. For supper, steak, gravy, mashed potatoes, margarine, canned corn, diet lemonade, and low-fat chocolate milk with fiber. Daily snacks provided were baked potato chips, dry roasted peanuts, cheese and peanut butter sandwich biscuits, Goldfish Crackers, and apple sauce.

A "minimally processed" day offered a breakfast of Greek yogurt parfait with strawberries, bananas, walnuts, and olive oil along with apple slices dipped in freshly squeezed lemon juice. Lunch consisted of chicken breast, bulgur, sunflower seeds, apple slices, and grapes accompanied by a spinach salad with a vinaigrette of olive oil, fresh squeezed lemon juice, apple cider vinegar, and ground mustard seed. Dinner was roast beef and a side of rice pilaf made with basmati rice, garlic, onions, sweet peppers, and olive oil. There was steamed broccoli as well as a salad of lettuce, tomatoes, cucumbers, and orange slices with balsamic vinaigrette. Daily snacks were oranges, apples, raisins, almonds, and walnuts.

The results were stunning. On the ultra-processed diet, people chose to eat about 500 calories more per day, resulting in a weight gain of two pounds over two weeks! For the first time, a study actually showed that processed foods were not only associated with but were the cause of weight gain! The next challenge is to find out why. That will take more research. Nutritional research, though, can be likened to a race towards a finish line that always seems to stay just out of reach.

I have spent much of my professional life reaching for that elusive finish line. When it comes to diet, I have read thousands of papers, waded through umpteen books, sat through countless lectures, had numerous discussions with experts, conducted surveys, tried out a myriad recipes, sampled an array of foods, and researched answers to innumerable questions posed to me about food. Recently, however, I had a question that gave me pause: "So, based on everything you have learned, what do you actually eat?"

Hmmm . . . first, I should say I obviously recognize that what we eat matters, otherwise this book would not exist. However, I also think that there is more to life than evaluating every morsel we eat or every drop we drink as being "good" or "bad" for us. There are healthy diets and unhealthy diets, but no one food should be singled out as a devil or an angel. And remember that while food is important, it is by no means the only factor governing health. Genetics, activity level, microbes, environmental contaminants, air pollution, sun exposure, and plain luck are all determinants. But food gets a great deal of attention because while we may not be able to control many other factors, we can control what we plunk into our mouth. So here we go.

I try to make sure I eat at least five, but usually a couple more, servings of fruits and vegetables a day and emphasize whole grains when it comes to bread. Sourdough is my favorite. For breakfast, I usually have berries with a high-fiber cereal, oatmeal, or unflavored yogurt. I add a spoonful of ground flaxseed unless I forget. I don't shy away from eggs or dairy, and I love potatoes in any form, although I limit French fries and other fried foods but not neurotically. I wish I could eat fish, but due to an allergy, I can't. So when it comes to meat, which I don't eat every day, it is mostly chicken, but I also like the occasional tasty hamburger and do not rule out eating a steak now and then. I restrict processed meats, only eat hot dogs at hockey games, and virtually never consume a soft drink. I rarely drink alcohol, but I enjoy coffee. I snack on fruit or nuts and may even indulge in pastry, but only if it is high quality. Twinkies or donuts need not apply for entry. I don't take any dietary supplements, I ignore "gluten-free," "non-GMO," and especially "chemical-free" claims. I don't look for "organic," but if it is the same price as conventional, I'll buy it. I try to keep added sugar to below 40 grams a day, and I watch my salt intake, restricting daily sodium to under two grams. There it is. Of course, it goes without saying that you can take all of this with a large grain of salt. Iodized.

INDEX

At ECW Press, we want you to enjoy this book in whatever format you like, whenever you like. Leave your print book at home and take the eBook to go! Purchase the print edition and receive the eBook free. Just send an email to ebook@ecwpress.com and include:

- the book title
- the name of the store where you purchased it
- your receipt number
- your preference of file type: PDF or ePub

A real person will respond to your email with your eBook attached. And thanks for supporting an independently owned Canadian publisher with your purchase!